AF560068

Instructional Video Production

NIPA® GENX ELECTRONIC RESOURCES & SOLUTIONS P. LTD.
New Delhi-110 034

About the Authors

Dr. P. Jaisridhar is a distinguished Assistant Professor at the ICAR – Krishi Vigyan Kendra, Tamil Nadu Agricultural University (TNAU), The Nilgiris with a specialization in Agricultural Extension. With a passion for teaching, Dr. Jaisridhar has delivered a variety of undergraduate and postgraduate courses, organized educational tours, and guided numerous students, including postgraduate research. His scholarly contributions include nine books several chapters, and a plethora of research papers and e-learning resources. Recognized for his research excellence, he has garnered multiple international and national awards, latest being Manak Veer Award from the Bureau of Indian Standards. His dedication to agricultural innovation and extension is reflected in his involvement in numerous externally funded projects and his leadership in organizing training programs and workshops for farmers and students alike. His role in institutional development at TNAU underscores his commitment to advancing agricultural education and practice.

Ms. I. Ponsneka is a dedicated Post Graduate student at Tamil Nadu Agricultural University (TNAU), who brings fresh perspectives and innovative ideas to the field of agricultural extension. Guided by Dr. P. Jaisridhar, her PG Thesis focussed on Role of Institutions in Mitigating Climate Change. Her collaboration with the author on "Instructional Video Production" marks a significant milestone in her academic journey. As the second author, Ms. Ponsneka contributes her research acumen and firsthand insights into the resources & software available for video and audio editing, production of films etc.

This book, Instructional Video Production, represents a comprehensive roadmap for mastering the art and science of creating professional-quality instructional videos. It serves as a practical guide and a learning resource that bridges the gap between theory and hands-on application in video production.

Instructional Video Production

P. Jaisridhar
I. Ponsneka

NIPA® GENX ELECTRONIC RESOURCES & SOLUTIONS P. LTD.
New Delhi-110 034

NIPA® GENX ELECTRONIC RESOURCES & SOLUTIONS P. LTD.

101,103, Vikas Surya Plaza, CU Block
L.S.C. Market, Pitam Pura, New Delhi-110 034
Ph : +91-11-43860225, Mob.: +91 9717133558, 9540816132
E-mail: newindiapublishingagency@gmail.com
Website: www.nipaersources.com

Print ISBN: 978-93-58873-23-8
ebook ISBN: 978-93-58872-89-7

Composed and Designed by NIPA®.

Preface

The art and science of instructional video production have undergone a significant transformation, driven by technological advancements and the growing demand for engaging, high-quality visual content. Whether in education, media, marketing, or entertainment, videos serve as a powerful medium for communication and learning. This book, Instructional Video Production, is a comprehensive guide designed to empower individuals and professionals to master the intricacies of video creation, editing, and presentation.

This book caters to a diverse audience, ranging from educators and students to aspiring filmmakers and content creators. It takes a practical approach, providing step-by-step instructions, hands-on exercises, and detailed explanations of the tools and techniques essential for producing compelling instructional videos.

The chapters are organized to guide readers through the entire production process, beginning with the fundamentals of handling DSLR and video cameras and culminating in advanced topics such as video editing with artificial intelligence and open-source broadcasting software.

Key highlights of the book include:

1. **Hands-On Practice with Equipment:** Readers are introduced to DSLR and video cameras, gaining practical experience in shooting high-quality videos.
2. **Mastering Software Tools:** The book covers open-source video and audio editing tools, as well as professional software like Adobe Photoshop, empowering readers to create polished videos without expensive resources.
3. **Smartphone Editing Techniques:** With the ubiquity of smartphones, the book explores mobile apps for video editing, synchronization, and exporting, making video production accessible to all.
4. **Creative Design and Animation:** Special attention is given to designing video titles, creating animations, and using tools like GIMP, Blender, and Google SketchUp.

5. **Advanced Editing Techniques:** Readers delve into advanced editing, voice synchronization, and integrating graphics and animation into videos.
6. **Cutting-Edge Technologies:** The book explores innovative tools such as AI-driven video creation and open broadcasting software to keep readers ahead in the field.
7. **Practical Application:** Chapters on recording techniques, offline and online editing, and final presentation ensure a well-rounded understanding of the production process.

Each chapter is designed to be standalone yet interconnected, offering readers the flexibility to navigate the topics based on their interests and skill levels. The inclusion of open-source tools highlights the book's commitment to making video production accessible and affordable.

The overarching aim of Instructional Video Production is to provide a solid foundation in video creation while inspiring creativity and innovation. By blending technical knowledge with practical insights, this book equips readers to harness the full potential of modern video production tools and techniques, ensuring they can produce impactful and professional instructional videos.

I hope this book serves as a valuable resource for you, whether you are embarking on your journey as a video creator or looking to enhance your existing skills.

Authors

Contents

1

Hands on Experience of DSLR and Video Camera

A DSLR, or "Digital Single Lens Reflex" camera, uses a mirror mechanism to direct light from the lens to an optical viewfinder, allowing the photographer to see exactly what the lens sees. When the shutter is activated, the mirror moves aside, allowing light to reach the image sensor, capturing the photograph.

While single-lens reflex cameras have been around in various forms since the 19th century, initially relying on film, the first commercially available digital SLR with an image sensor emerged in 1991. Unlike point-and-shoot or phone cameras, DSLR cameras allow for interchangeable lenses, offering greater flexibility.

Working Principles of DSLR Camera

When using a DSLR camera, the image seen through the viewfinder at the back of the camera is the actual view through the lens. This ensures you are previewing the precise scene you intend to capture. Light from the scene enters through the lens and hits a reflex mirror positioned at a 45-degree angle inside the camera. This mirror directs the light vertically to an optical component known as a "pentaprism." The pent prism then reorients the light horizontally by passing it through two internal mirrors, delivering the final image into the viewfinder.

When the shutter button is pressed to take a picture, the reflex mirror swings up, blocking the vertical light path, and allowing light to pass directly through. The shutter then opens, allowing light to reach the image sensor. The shutter stays open for the required time to expose the image sensor, after which it closes, and the reflex mirror returns to its original 45-degree angle to redirect light back to the viewfinder. Once the image is captured, the camera's processor takes over. It processes the data from the image sensor, formats it, and saves it to a memory card. This entire process is quick, with some professional DSLRs capable of capturing over 11 images per second.

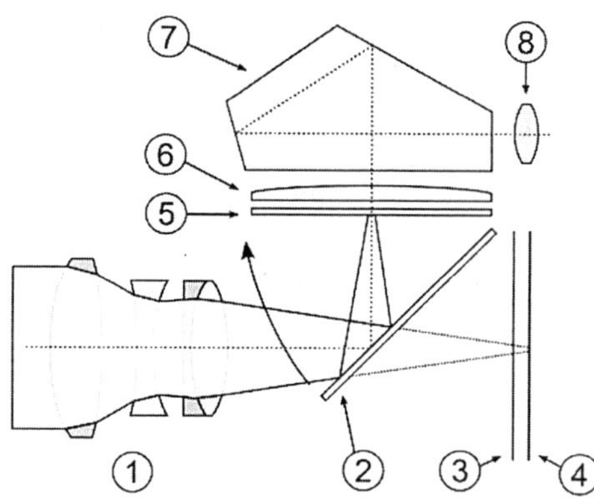

① Lens
② Reflex Mirror
③ Shutter
④ Image Sensor
⑤ Focusing Screen
⑥ Condenser Lens
⑦ Pentaprism
⑧ Eyepiece / Viewfinder

Source: Wikipedia

Examples of Recent DSLR Models

Over the years, numerous DSLR models have been produced, with some standing out as remarkable choices. One highly regarded model is the **Nikon D850**, equipped with a 45.7MP BSI sensor that provides exceptional dynamic range and fine detail. Although the D850's autofocus is surpassed by models like the D500 and D5/D6, it still performs reliably, making it one of the top DSLRs available.

In Canon's lineup, the **Canon 5D Mark IV** parallels the D850 in excellence. This model offers a sophisticated autofocus system that handles fast-action photography with ease. Canon's dual-pixel AF technology, built into the 5D Mark IV, also enables excellent video capabilities—an area where Nikon models typically fall short.

For Pentax enthusiasts, the **Pentax K1 Mark II** represents a significant development as the first full-frame DSLR from Pentax. While its autofocus may not match the high-performance systems of Canon and Nikon, the K1 Mark II excels in landscape photography. It includes a unique astrotracer feature, ideal for capturing detailed Milky Way images, setting it apart in the realm of night photography.

Difference Between DSLRs and Mirror less Cameras

Today, many new camera models are mirror less, meaning they lack a defining feature of the DSLR: the mirror.

In DSLRs, a mirror and prism system redirects light from the lens to an optical viewfinder, enabling you to see the scene exactly as it appears through the lens. Mirrorless cameras, by contrast, do not have this mirror and prism mechanism and therefore do not use an optical viewfinder. Instead, mirrorless cameras rely on light directly hitting the digital sensor to generate a preview of the scene, which is displayed on an electronic viewfinder (EVF) rather than an optical one.

DSLRs, however, often include a "Live View" mode, which raises the mirror permanently and shows a preview image on the rear screen, allowing the DSLR to function similarly to a mirrorless camera in this mode. Unlike mirrorless cameras, though, this preview is not available through an EVF.

Both DSLRs and mirrorless cameras are suitable for almost any type of photography. Here are some primary differences in how each type is typically used:

Advantages of DSLRs vs. Mirrorless Cameras

DSLR Advantages

- **Longer Battery Life**: DSLRs typically offer extended battery performance compared to mirrorless models.
- **Optical Viewfinder with No Delay**: The optical viewfinder provides a real-time, lag-free view of the scene.
- **Availability of Used Models**: DSLRs have a larger market for used models, often at affordable prices.
- **Compatibility with Older Lenses**: Many DSLRs support a wider range of older lenses without adapters.

Mirrorless Advantages

- **Enhanced Video Capabilities**: Mirrorless cameras often feature advanced video functionalities.
- **Advanced Autofocus**: Many mirrorless models are equipped with sophisticated autofocus systems.
- **WYSIWYG Viewfinder**: Mirrorless electronic viewfinders display exactly how settings will affect the final image, providing a "What You See Is What You Get" (WYSIWYG) view.
- **Continued Manufacturer Support**: Mirrorless technology is rapidly evolving, with active support and frequent updates from manufacturers.

Disadvantages of DSLRs

While digital cameras allow immediate viewing of images, DSLR viewfinders do not always display precisely what the image sensor captures. This is due to DSLRs relying on the mirror for aspects of focusing, which can result in a slight mismatch between what is seen in the viewfinder and the actual captured image. As a result, photographers may not see the exact exposure of their shot until editing.

In contrast, mirrorless cameras display the live image directly from the sensor, providing a more accurate preview of exposure, making it easier to adjust settings in real-time. With DSLRs, although adjustments can be made on the spot, photographers may face more editing challenges afterward due to exposure differences between the viewfinder and the sensor.

Editing images from a DSLR can require some practice. Photographer Stephen Klise notes that switching to digital presented a learning curve, as light and color interact differently than in traditional photography, with prominent reds being particularly noticeable.

Software like Adobe Photoshop Lightroom offers extensive tools to address DSLR-specific color corrections. Some helpful Lightroom features include:

- **Adjusting Light and Color**: Use the Light panel to fine-tune white balance, saturation, and other color settings.
- **Removing Photo Tints**: Quickly remove color casts caused by lighting conditions.
- **Enhancing Color Intensity**: Adjust the Vibrance and Saturation sliders to make colors in your images stand out.

These tools provide flexibility in handling DSLR-specific adjustments and optimizing image quality.

Why Photographers Choose DSLR Cameras

Photographers often favor DSLR cameras for their durability, versatility with a wide range of lenses and attachments, long battery life, high shooting speeds, and advanced autofocus capabilities. However, DSLRs have a limitation: there can be a difference between what is visible in the viewfinder and the actual exposure captured, an issue typically avoided with mirrorless cameras.

Both DSLRs and mirrorless cameras support continuous shooting (burst mode) and offer image stabilization features. However, mirrorless cameras have an edge in video recording. The DSLR's mirror system complicates video autofocus, making mirrorless models more suitable for capturing full HD video. Additionally, mirrorless cameras are generally lighter and more compact, as they only require space for a sensor rather than a mirror system. Music photographer Chad Wadsworth, an early user of mirrorless technology, highlights these advantages for photographers seeking portability and ease of video capture.

Choosing between a DSLR and a mirrorless camera depends on your photography subject and shooting conditions. Different DSLR models and lens options offer unique advantages, and understanding these distinctions can help you find the best camera for your needs.

Video Camera: Overview and Components

A video camera is an optical device that captures videos, unlike a movie camera, which records images on film. Initially developed for the television industry, video cameras are now widely utilized for various purposes.

Modes of Use

Video cameras operate primarily in two modes:

1. **Live Television Mode**: This mode captures real-time images, transmitting them directly to a screen for immediate viewing. Initially common in broadcasting, this mode now finds frequent application in security, military, and industrial operations that require surreptitious or remote observation.
2. **Recorded Mode**: In this mode, images are stored on a device for later archiving or processing. Early video recordings were stored on videotape, but this format has largely been replaced by optical discs, hard drives, and flash memory. Recorded video is often used in television production, surveillance, and monitoring, where unattended recording and later analysis are necessary.

Types and Uses of Video Cameras

Modern video cameras have evolved into various designs to meet specific needs:

- **Professional Video Cameras**: Used in television production, these cameras may be based in studios or used in the field for electronic field production (EFP). They offer detailed manual controls, often bypassing automation, and usually utilize three sensors to capture red, green, and blue separately.
- **Camcorders**: Combining a camera and recording device into one unit, camcorders are mobile and widely used for television production, home movies, and electronic news gathering (ENG). Since the digital transition, most cameras with built-in recording media qualify as camcorders. Action cameras, a sub-type, often feature 360° recording capabilities.
- **Closed-Circuit Television (CCTV) Cameras**: These pan–tilt–zoom (PTZ) cameras are primarily used for security, surveillance, and monitoring. They are compact, easily hidden, and designed for unattended operation, with specialized industrial variants suited for hazardous environments such as high-radiation or chemically volatile areas.

- **Webcams**: These are designed to stream live video directly to a computer.
- **Smartphone Cameras**: Many smartphones now feature built-in video cameras, with high-end models capable of recording in 4K resolution.
- **Scientific Cameras**: Specially engineered for research applications, these cameras are often used in space exploration, artificial intelligence, robotics, and medical research. They may capture non-visible radiation, including infrared for night vision and X-ray for medical imaging.

Historical Developments

The evolution of video cameras began with early mechanical systems like the Nipkow disk used in experimental broadcasts during the 1910s–1930s. By the 1930s, all-electronic video camera tubes, such as Vladimir Zworykin's Iconoscope and Philo Farnsworth's image dissector, replaced the Nipkow disk. These systems dominated until the 1980s when solid-state image sensors, including charge-coupled devices (CCD) and later CMOS active-pixel sensors, solved many issues of the tube-based systems, such as image burn-in and streaking.

The innovation of **solid-state image sensors** originated from metal–oxide–semiconductor (MOS) technology. The first significant advancement, the charge-coupled device (CCD), was invented at Bell Labs in 1969. This was followed by the development of CMOS sensors at NASA's Jet Propulsion Laboratory in 1993, enhancing the practical application of digital video workflows by eliminating the need for analog-to-digital conversion.

Digital Transition and Compression

Digital video cameras gained prominence due to advancements in **video compression technology**, particularly the discrete cosine transform (DCT) algorithm. Developed in 1972, DCT-based compression paved the way for standards like H.26x and MPEG, introduced from 1988 onwards. This breakthrough enabled video cameras to manage the high memory and bandwidth requirements of digital video, significantly benefiting digital television and video capture.

Professional Video Equipment Evolution

With digital capture, the line between professional video cameras and movie cameras has blurred, as both now use similar mechanisms. By the early 21st century, most video cameras had become digital. The early experimental recording of video, notably John Logie Baird's Phonovision discs in 1927, was succeeded by tape-based recording technologies like Ampex's Quadruplex videotape in 1956. Portable recording systems, beginning with Sony's DV-2400 in 1967, evolved into camcorders like the Betacam system in 1981, where the recorder and camera were integrated.

Lens and Mount Options

Video cameras may come with built-in lenses or use interchangeable lenses connected via specialized mounts. Mounts such as Panavision PV and Arri PL are suited to movie cameras, while Canon EF and Sony E originate from still photography. Additionally, CCTV applications have unique mounts like the S-mount for specific uses.

Through various technological innovations, video cameras have diversified and adapted, making them indispensable across fields ranging from entertainment and media to research, security, and personal use.

Professional Video Camera

A professional video camera, often referred to as a television camera, is a sophisticated electronic device designed to capture high-quality moving images. Initially developed for television studios and outside broadcast units, professional video cameras have evolved to be used across various fields, including music video production, direct-to-video movies, corporate and educational videos, and even wedding videography. Since the 2000s, most professional video cameras have transitioned from analog to digital formats, facilitating high-definition (HD) recording and bringing their quality closer to traditional 35mm film cameras. HDTV cameras, specifically designed for broadcast television, news, sports, events, and reality television, are commonly classified as professional video cameras. The line between professional video and movie cameras has blurred with advancements in HD digital video cameras, which now feature sensor sizes, dynamic range, and color rendition capabilities comparable to film.

Professional video cameras offer both live broadcast capabilities and high-quality recording for post-production, typically needing minimal color and exposure adjustments. In contrast, digital movie cameras, often used in scripted television and movies, are designed for recordings that require more comprehensive color correction in post-production.

Historical Development of Professional Video Cameras

1. **Early Beginnings (1926–1933)**: Early "video cameras" were mechanical scanners using a flying spot and disk mechanism, a primitive form of image capture.
2. **Advancements in Electronic Cameras (1936–1950)**: RCA introduced the iconoscope camera in 1936. The Vidicon camera tube followed in 1950, allowing for more compact camera designs.
3. **Color Broadcasting and Enhanced Image Sensors (1950s–1970s)**: RCA's TK-40 was the first color broadcast camera in 1953. As technology advanced, the introduction of devices like the Image Orthicon tube and Plumbicon improved image clarity. The TK-47, RCA's high-end tube camera from 1978, marked the height of tube camera technology before CCD sensors took over.
4. **Modern Digital Transition (1980s–2000s)**: By the 1980s, solid-state sensors such as charge-coupled devices (CCD) were introduced, eventually leading to today's CMOS sensors. This technology allowed cameras to avoid common tube problems, making digital workflows more practical.

Camera Design and Operation

Professional video cameras generally utilize an optical prism block that sits directly behind the lens. This prism separates light into the three primary colors—red, green, and blue—and directs each to a separate CCD or CMOS sensor. These sensors create a digital signal that is either broadcast live or stored for post-production editing. In both single-sensor (using a Bayer filter) and three-sensor designs, the weak signal from the sensors is amplified and then encoded for use in various outputs, such as analog or digital signals. Analog outputs often include a composite video signal or component video output, whereas digital outputs are increasingly preferred for modern broadcast and recording.

In television studios, professional cameras typically stand on pedestals—hydraulic or pneumatic mechanisms that allow operators to adjust the height and position of the camera easily. The multiple-camera setup is managed from a camera control unit (CCU), which regulates the signals and provides consistency in the broadcast. These units connect to the camera via triaxial cables, fiber optics, or multicore cables.

When used in outdoor broadcasts, these cameras are often mounted on tripods or dollies. Studio cameras can also be handheld by detaching from the pedestal and swapping out the lens for a smaller, portable option. Tally lights signal to operators and subjects that the camera is currently live.

Types of Professional Video Cameras

1. **ENG (Electronic News Gathering) Cameras**: These are portable cameras intended for field reporting, such as news gathering. ENG cameras are typically larger and heavier to stabilize hand-held shots and include shoulder supports for ease of use. They are equipped with professional-grade sensors and manual controls, including settings for white balance, gain, and color bars.
2. **Studio Cameras**: Stationary and often mounted on pedestals or tripods, studio cameras are ideal for multi-camera setups in controlled environments. They feature advanced manual controls and are connected to CCUs for centralized control of broadcast settings.
3. **Remote Cameras and Block Cameras**: Used for capturing footage in hard-to-reach locations, remote cameras are typically controlled from a distance and are mounted on pan-and-tilt heads or robotic arms. Block cameras, named for their compact size, are useful in settings where space is limited and can be remotely operated.
4. **Lipstick Cameras**: Small and lightweight, lipstick cameras have a sensor block and lens setup resembling a lipstick tube, suitable for capturing video in tight or mobile spaces, such as race cars. These cameras are often attached to boom poles or mounted for steady, fixed shots.

Recording Formats and Media

Professional video cameras initially recorded to analog media, but with the transition to digital technology, formats like Betacam, DVCPRO, and direct-to-disk storage became standard. Today, flash memory has become the dominant storage medium, supporting high-quality, compressed digital files with minimal degradation. Additionally, "bars and tone," a standard reference signal for calibration, helps in post-production to ensure accurate color and sound quality. The development of these video systems from analog to digital storage has enhanced the accessibility and versatility of video production in numerous fields.

Hands-On Camera Techniques for Visual Storytelling

Pre-Shoot Planning

Effective camera work begins before you even arrive on set. Planning your shots enhances storytelling by setting clear visual objectives. Discuss with the team or reporter the ideal shots to capture, such as landscapes, close-ups, or dynamic actions, aligning these visuals with the story's theme. Having a mental or written shot list helps keep your vision clear, though flexibility on location is also essential.

Preparing Your Equipment

Before beginning a shoot, insert a new recording tape or check your digital storage, allowing 30 seconds to roll to ensure everything is functioning correctly. Test audio levels and, ideally, wear headphones throughout the shoot to monitor quality. Create a pre-shoot checklist covering essentials like batteries, lens cleaning, and storage, which will eventually become second nature.

Selective Shooting

To save time during editing, shoot selectively. Capture only what you need, avoiding unnecessary footage, which can clutter the editing process. Start thinking critically about each shot as soon as you begin recording, and develop habits that will allow you to capture compelling footage from the outset.

Quiet on Set!

When recording, it's crucial to maintain silence. Ambient sounds such as cars or wind may enhance realism, but background conversations can ruin audio quality, making editing difficult. Capture shots with slight buffers—before and after the action—to provide flexibility for trimming in post-production.

Minimal Zooming and Panning

Avoid overusing zoom and pan functions, as they can detract from the composition. Instead, frame your subject carefully and hold the shot steady. Only zoom or pan when it serves the story, such as transitioning between subjects or highlighting a detail, and always remember to keep shots steady and composed.

Establishing Shots and Visual Sequences

Effective sequences build a narrative. For example, if you're filming at a gas station, begin with a wide shot to establish the scene, followed by medium and close-up shots focusing on specific details, like a hand pulling the gas nozzle. These sequences help visually narrate the story in a way that feels dynamic and engaging.

Varying Angles and Perspectives

To make footage more interesting, vary angles and perspectives, especially for online platforms where close-up details resonate more. Avoid shooting everything from eye level; experiment with high, low, and angled shots for greater visual impact. However, stay mindful of authenticity—don't manipulate shots to suggest a false reality.

Audio Layering and Composition

Effective audio layering adds depth, so consider capturing different sound types and mixing them to create a richer experience. Incorporate ambient noise to reflect the location, and position subjects within the frame using the "rule of thirds" for a balanced composition. When interviewing, ensure the camera is at eye level to avoid excessive empty space.

Interview Techniques

When conducting interviews, frame the subject's face with their eyes aligned to the upper third of the screen. Avoid direct eye contact with the camera, as it can be distracting. If working alone, maintain eye contact while keeping one ear on the audio to monitor sound levels.

Using Microphones Effectively

When using a lapel mic, ensure it's positioned subtly but directly in line with the speaker's voice. With handheld microphones, keep them out of the shot but directed toward the audio source. Accurate microphone placement contributes significantly to high-quality sound.

Capture Reverse and Cutaway Shots

Reverse and cutaway shots help provide smooth transitions and pacing within a segment. For interviews, capture a few seconds over the interviewer's shoulder, followed by shots from the opposite angle. This technique allows flexibility in the editing room and adheres to the "180-degree rule" to maintain spatial consistency.

Lighting Techniques

Three-point lighting, consisting of a key light, fill light, and backlight, enhances subject visibility and adds depth. When time permits, set up this traditional lighting arrangement to achieve professional results. In fast-paced settings, improvisation with available light sources is often necessary.

Developing Creative Control

Like mastering classical techniques in painting before experimenting, mastering camera fundamentals is crucial. Build a strong foundation by practicing stable framing, optimal lighting, and audio clarity. Once you're comfortable, experiment with angles and sequences to develop a unique style that enhances storytelling.

Quality and Storytelling

When producing videos for digital platforms, quality remains essential despite faster production demands. Push for footage that enhances the overall narrative, adding new insights beyond what text can convey. Video should not merely accompany the story but provide a compelling, informative layer that text alone cannot achieve.

In summary, good camera work is more than technical proficiency—it is a blend of planning, precision, and creativity that together create a compelling visual narrative.

2

Open Source Software for Video Editing

Introduction to Open Source Software in Video Editing

In the digital age, video editing has become an essential skill, both for professionals in the media industry and for hobbyists. Open source software, with its flexibility, affordability, and community-driven support, has emerged as a popular choice for video editing. Unlike proprietary software, open source software is freely available and can be modified and shared by users, allowing for constant updates and improvements.

Open source video editing tools provide a wide array of features, from basic trimming and cutting to advanced color grading, visual effects, and audio editing. These tools not only serve as cost-effective alternatives to commercial software but also encourage community involvement and learning.

Key Features of Open Source Video Editing Software

Open source video editors often include many features found in professional-grade software, such as:

- **Non-linear Editing (NLE)**: This allows users to access and edit any part of the video file in a non-sequential manner, giving greater flexibility.
- **Multi-track Editing**: Users can work with multiple video and audio layers, making it easier to combine elements for complex edits.
- **Visual Effects and Transitions**: Many open source editors support a variety of effects and transitions to enhance visual storytelling.
- **Audio Editing Capabilities**: These programs often provide basic audio editing tools and support for multi-track audio, essential for producing high-quality sound.
- **Color Grading and Correction**: Some open source editors allow users to adjust brightness, contrast, and color, ensuring a polished, professional look.

This chapter covers popular open source video editing software and their applications, advantages, and limitations. By understanding these tools, readers can make informed choices about which software best suits their editing needs.

Popular Open Source Video Editing Software

1. Kdenlive

Kdenlive, short for KDE Non-Linear Video Editor, is a powerful open source video editor available for Linux, macOS, and Windows. It is known for its robust set of features, including multi-track video editing, support for various video formats, and advanced color correction tools.

Source: kdenlive.org

- **Features**
  - Multi-track editing and real-time effects.
  - Support for a wide variety of video and audio formats.
  - Advanced color grading tools.
  - A customizable interface and support for plugins.
- **Applications**: Kdenlive is ideal for intermediate to advanced users who need a professional-grade editor without the cost of proprietary software. Its extensive community support and documentation make it accessible, while its advanced features cater to high-level editing projects.
- **Limitations**: Although powerful, Kdenlive may sometimes experience stability issues, particularly with high-definition or high-bitrate files. Its learning curve may also be steep for beginners, though its large user community offers extensive tutorials and troubleshooting resources.

2. Blender

Primarily known as a 3D animation software, Blender also includes a powerful video editor. It is unique in the open source community for its combination of 3D modeling, animation, and video editing capabilities, making it an all-in-one tool for multimedia projects.

Source: blender.org

- **Features**
  - A video sequence editor with transitions, speed control, and adjustment layers.
  - Full 3D modeling and animation tools for integrating animations directly into video projects.
  - Advanced features for visual effects (VFX) and motion tracking.
  - Python scripting for customization.
- **Applications**: Blender is suited for users who need both video editing and animation capabilities in one package. It's widely used in the film and game development industries, making it a versatile choice for those looking to create complex, professional-quality media.
- **Limitations**: Blender's comprehensive toolset can be overwhelming for new users, especially those who are not familiar with 3D animation. The interface may also be challenging to navigate for those who only need basic video editing features.

3. Shotcut

Shotcut is a cross-platform open source video editor that emphasizes simplicity and ease of use. It's known for its straightforward, user-friendly interface and robust editing features, making it an excellent choice for beginners.

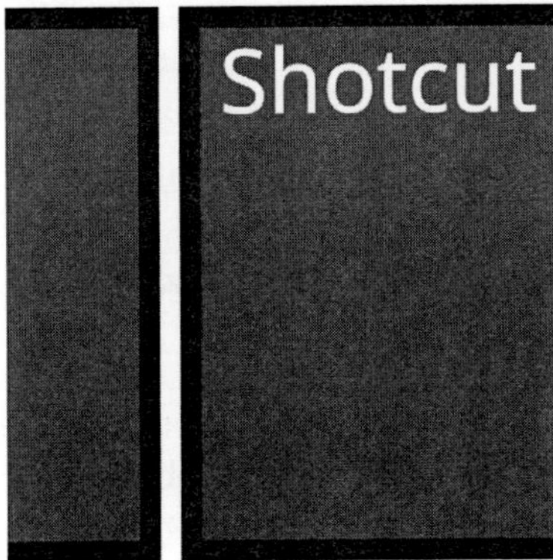

Source: shotcut.org

- **Features**
  - Support for a wide range of formats thanks to FFmpeg integration.
  - A variety of video and audio filters.
  - Frame-accurate trimming and editing.
  - Native timeline editing without the need for importing.

- **Applications**: Shotcut is ideal for beginners and casual users who need reliable video editing capabilities without the complexity of advanced features. It's frequently used for simple video projects, such as creating YouTube content or editing short films.
- **Limitations**: While Shotcut is beginner-friendly, it lacks some advanced features found in software like Blender and Kdenlive. Users requiring complex visual effects or advanced color grading may need to look elsewhere.

4. OpenShot

OpenShot is an open source video editor known for its intuitive design and ease of use, making it accessible to beginners while offering features useful for more complex projects.

Source: openshot.org

- **Features**
  - Cross-platform compatibility.
  - Multi-track editing, unlimited layers, and keyframe animations.
  - Built-in transitions and video effects.
  - Simple drag-and-drop interface.
- **Applications**: OpenShot is frequently used for educational purposes and small projects where ease of use is a priority. It's also popular among YouTube creators and students due to its simplicity and accessible design.
- **Limitations**: OpenShot can occasionally experience performance issues with high-resolution videos, and its advanced capabilities may be limited compared to Kdenlive or Blender. It's best suited for simpler editing tasks and may not meet the needs of professionals working on complex projects.

5. DaVinci Resolve (Free Version)

Overview: While DaVinci Resolve is not entirely open-source, its free version is one of the most advanced video editors available, and it includes a comprehensive set of tools for professional editing, color grading, VFX, and audio post-production.

Source: https://www.blackmagicdesign.com/in/products/davinciresolve/studio

- **Key Features**
  - Professional-grade color correction and grading tools.
  - Fusion for advanced visual effects and motion graphics.
  - Fairlight audio editing panel for high-end audio editing and mixing.
  - Collaboration features for team-based editing environments.
- **Supported Platforms**: Windows, macOS, and Linux.
- **Use Case Example**: DaVinci Resolve is frequently used in film and TV production, making it ideal for editors who want professional-grade editing tools, especially for complex projects involving advanced color grading and sound design.

6. Avidemux

Overview: Avidemux is a simple, lightweight video editor geared towards quick tasks like trimming, encoding, and basic filtering. It lacks a timeline but is valuable for straightforward editing tasks and conversions.

Source: https://avidemux.sourceforge.net/

- **Key Features**
  - Simple cutting, filtering, and encoding for fast edits.
  - Support for multiple file types, including AVI, DVD-compatible MPEG files, MP4, and ASF.
  - Basic filter options for resizing, deinterlacing, and color correction.
- **Supported Platforms**: Windows, macOS, and Linux.
- **Use Case Example**: Avidemux is ideal for editors who need a straightforward tool for quick edits or format conversions without the complexity of a full editing suite.

7. Olive

Overview: Olive is an open-source video editor that aims to provide professional-grade editing capabilities. Although still in development, it has garnered attention for its sleek design and advanced features similar to those found in commercial software.

Source: https://www.olivevideoeditor.org/about

- **Key Features**
  - Intuitive interface with non-linear editing capabilities.
  - Advanced features for color correction, transitions, and effects.
  - Real-time playback and advanced rendering engine for smoother workflow.
  - Regular updates as an active open-source project.
- **Supported Platforms**: Windows, macOS, and Linux.
- **Use Case Example**: Olive is suited for users looking for a free video editor that may eventually rival industry-standard software in features and functionality, ideal for intermediate to advanced projects.

8. Flowblade

Overview: Flowblade is a free and open-source video editor for Linux. Known for its stability and responsive performance, it's an excellent choice for Linux users seeking a reliable editor for complex projects.

Source: https://jliljebl.github.io/flowblade/

- **Key Features**
  - Supports multi-track editing with frame-accurate trimming and audio mixing.
  - Includes compositing tools, keyframes, and a range of transitions and effects.
  - Proxy editing support for editing high-resolution files smoothly on slower systems.
  - Compatible with a wide array of file formats through FFmpeg.
- **Supported Platforms**: Linux.
- **Use Case Example**: Flowblade is often used by Linux-based documentary creators, indie filmmakers, and video bloggers who need a powerful editor with reliable performance on Linux.

Advantages of Open Source Video Editing Software

1. **Cost-Effectiveness**: Open source software is free to use, which makes it accessible for individuals and organizations with limited budgets.
2. **Community Support**: Open source platforms benefit from active communities that provide support, share plugins, and continuously improve the software.
3. **Flexibility and Customization**: Many open source video editors are highly customizable, allowing users to add plugins or modify the code to better meet their needs.
4. **Cross-Platform Availability**: Most open source editors are compatible with Linux, Windows, and macOS, making them accessible to users across different systems.

Limitations of Open Source Video Editing Software

1. **Learning Curve**: Some open source software, such as Blender, can have a steep learning curve due to its extensive features.
2. **Stability Issues**: Open source software may sometimes lack the stability of commercial software, leading to crashes during high-demand tasks.
3. **Fewer High-End Features**: While powerful, open source tools may not always match the specialized features of proprietary software like Adobe Premiere Pro or Final Cut Pro.

Conclusion

Open source video editing software offers an accessible, cost-effective solution for both beginners and professionals. With options ranging from simple editors like Shotcut and OpenShot to advanced tools like Blender and Kdenlive, there is software available to meet a wide array of editing needs. The open source community also fosters learning and creativity, allowing users to experiment and customize their software to suit specific project requirements. As digital media continues to evolve, open source video editors are likely to play a growing role in democratizing content creation.

3

Open Source Software for Audio Editing

Introduction to Open Source Audio Editing Software

The field of audio editing has seen remarkable advancements, driven by open-source software that democratizes access to high-quality audio production tools. Unlike proprietary software, open-source audio editors are freely available, and they allow users to access, modify, and enhance the source code to suit specific needs. Open-source tools are popular among audio engineers, musicians, content creators, and educators for their cost-effectiveness, flexibility, and ability to operate across multiple platforms.

This chapter explores popular open-source software options for audio editing, their features, advantages, limitations, and how they compare with proprietary solutions. Additionally, we discuss the technical principles underpinning audio editing, such as waveform editing, multi-track capabilities, effects processing, and export functionalities.

1. Fundamentals of Audio Editing

Before delving into specific open-source tools, it is essential to understand the core features required for audio editing:

- **Waveform Editing**: Displaying audio signals as waveforms allows for precise editing, such as trimming, splitting, and moving audio segments.
- **Multi-track Editing**: This feature enables the layering of audio tracks to create complex compositions, mix music, or align voiceovers with background tracks.
- **Effects and Processing**: Basic effects include equalization (EQ), reverb, compression, and normalization, which are essential for refining the audio quality.
- **Export Options**: A variety of export formats (MP3, WAV, FLAC, etc.) are needed for different purposes, from high-fidelity production to compressed web formats.

2. Overview of Open Source Audio Editing Software

2.1 Audacity

Audacity is one of the most popular open-source audio editing software tools, renowned for its user-friendly interface and versatility. Available on Windows, macOS, and Linux, Audacity is frequently used for tasks ranging from podcast production to basic music mixing.

Source: https://www.audacityteam.org/

- **Features**
  - **Multi-track editing**: Allows users to work with multiple audio tracks, making it ideal for creating layered soundscapes.
  - **Range of effects**: Includes built-in effects such as EQ, reverb, and noise reduction, with additional support for VST (Virtual Studio Technology) plugins.
  - **Spectrogram and frequency analysis**: A visual display for precise audio adjustments and corrections.
  - **File import and export**: Supports WAV, AIFF, MP3, and other formats, making it adaptable to different media needs.
- **Advantages**
  - Comprehensive audio processing tools suitable for beginners and experienced users.
  - Extensive community support and online tutorials.
- **Limitations**
  - Limited to destructive editing, meaning changes are directly applied to the audio file, unlike non-destructive editing in high-end proprietary software.

2.2 Ardour

Ardour is a digital audio workstation (DAW) tailored for audio engineers and professional users seeking advanced capabilities. Often compared to high-end

DAWs, Ardour's robust features cater to music production, film scoring, and live recording.

Source: https://ardour.org/

- **Features**
  - **Non-destructive editing**: Allows users to edit without altering the original audio, which is beneficial for large projects requiring multiple revisions.
  - **MIDI integration**: Supports MIDI editing, enabling users to integrate virtual instruments and external MIDI devices.
  - **Advanced effects**: Offers dynamic control with plugins for precise audio shaping.
  - **Automation tools**: Allows users to automate volume, panning, and other parameters for complex audio arrangements.
- **Advantages**
  - Suitable for high-level audio production, offering features comparable to paid DAWs.
  - Strong support for Linux users and compatibility with multiple audio plugins (VST, LV2, LADSPA).
- **Limitations**
  - Steeper learning curve, which may be challenging for beginners.
  - Linux-centric, with limited support for Windows and macOS.

2.3 LMMS (Linux Multimedia Studio)

Originally designed for Linux but now available on Windows and macOS, LMMS is an open-source audio editor and DAW focused on music production. It includes features suited for MIDI sequencing, making it a popular choice for musicians and electronic music creators.

Source: https://lmms.io/

- **Features**
  - **MIDI piano roll**: For creating, arranging, and modifying MIDI notes.
  - **Built-in synthesizers and samples**: Offers a wide range of sounds for music creation, including drum kits and virtual instruments.
  - **Automation editor**: Enables control over volume, pitch, and effects over time.
  - **Cross-platform support**: Projects can be saved and reopened across platforms, allowing for flexibility.
- **Advantages**
  - Excellent choice for electronic music production and MIDI-based projects.
  - User-friendly interface with an active community contributing plugins and samples.
- **Limitations**
  - Limited audio editing features compared to Audacity and Ardour, making it less suitable for complex audio processing.
  - Primarily oriented toward music creation, not general-purpose audio editing.

2.4 Ocenaudio

Ocenaudio is a lightweight, cross-platform audio editor with a straightforward interface, ideal for users who need basic audio editing without the complexities of a DAW.

Source: https://www.ocenaudio.com/

- **Features**
  - **Real-time preview of effects**: Users can hear the effects applied to audio in real-time, making adjustments easier.
  - **Multi-selection editing**: Allows users to select multiple parts of an audio file simultaneously, which speeds up editing.

- **Spectrogram view**: Detailed spectrogram display for analyzing audio frequencies.
- **Advantages**
  - User-friendly, suitable for quick edits and smaller audio projects.
  - Lightweight application with less demand on system resources.
- **Limitations**
  - Limited to basic editing tools, making it unsuitable for advanced multi-track editing or complex production projects.

2.4. Reaper (Not entirely open-source but has a substantial free evaluation period)

Overview: Reaper is a comprehensive DAW that offers a generous free evaluation period, although it's not fully open-source. Many users rely on it due to its customizable interface, scripting capabilities, and support for a wide range of plugins.

Source: https://www.reaper.fm/

Key Features

- Highly customizable interface and workflows
- Extensive third-party plugin support
- Built-in scripting for workflow automation

2.5. Waveform Free by Tracktion

Overview: Waveform Free, the free version of Tracktion's Waveform Pro, is available as a no-cost DAW with essential features for audio recording, mixing, and MIDI composition. While it's free and widely accessible, it is not fully open-source.

Source: https://www.tracktion.com/products/waveform-free

Key Features

- High-resolution mixing and recording tools
- MIDI editing and composition tools
- Compatible with VST and AU plugins

3. Technical Principles in Open Source Audio Editing

3.1 Waveform and Spectrogram Editing

Editing software like Audacity and Ocenaudio enables users to visualize audio as waveforms, making it easier to identify and adjust peaks, valleys, and irregularities in sound. Some software includes spectrogram editing, allowing frequency-based analysis critical for removing noise or fine-tuning vocal clarity.

3.2 Non-Destructive vs. Destructive Editing

Non-destructive editing, available in Ardour and other DAWs, maintains the original audio track, letting users experiment without affecting the raw data. Destructive editing, typically found in simpler editors, directly alters the audio file and is best suited for quick edits.

3.3 Real-Time Effects and Plugins

Open-source software often supports VST, LV2, and LADSPA plugins, which add various effects. Audacity, for example, supports a wide range of third-party plugins, enabling users to expand the software's capabilities for reverb, compression, and other effects.

3.4 Export Formats and Compatibility

Exporting in multiple formats (WAV, MP3, FLAC) is crucial, especially when projects are destined for platforms with unique requirements. Open-source audio editors like Audacity and LMMS offer extensive format options to meet these needs.

4. Comparison with Proprietary Software

While open-source tools are versatile, they often lack some advanced capabilities of proprietary software, like Adobe Audition or Avid Pro Tools.

Open-source solutions may have a limited set of native effects, fewer integrated loops, and less extensive customer support. However, their cost-effectiveness and flexibility, especially for users comfortable with the basics of audio editing, make them a valuable option.

Conclusion

Open-source audio editing software offers a powerful and accessible solution for users across all experience levels. With tools like Audacity, Ardour, LMMS, and Ocenaudio, users can access a range of features from simple edits to complex productions. Although there may be limitations compared to proprietary options, open-source software is continually evolving and gaining more support from communities and developers.

4

Video Editing Using Smart Phones in Mobile Apps

Introduction

In recent years, video editing has transformed with the rise of mobile apps, making high-quality video production accessible to anyone with a smartphone. The portability of smartphones and the power of mobile editing apps empower users to create, edit, and share videos from anywhere. This chapter explores the fundamentals of video editing on mobile devices, the range of available apps, their capabilities, best practices, and specific examples.

1. Advantages of Mobile Video Editing

1.1 Portability and Accessibility

Smartphones eliminate the need for complex and often expensive computer-based setups. Users can now shoot, edit, and publish videos directly from their device, making the process faster and more convenient.

1.2 Cost-Effectiveness

Most video editing apps on smartphones are either free or offer affordable subscriptions. This makes high-quality video production accessible to hobbyists, freelancers, and small businesses without the overhead of expensive editing software and hardware.

1.3 Instant Sharing Capabilities

Smartphones make it easy to share videos directly from the app to social media platforms, cloud storage, or messaging applications, facilitating instant engagement.

2. Key Features in Smartphone Video Editing Apps

Most mobile video editing apps provide several core features that cater to both beginners and more advanced editors. These features include:

- **Trimming and Cutting**: Basic edits to adjust the start and end points of clips.

- **Filters and Color Grading**: Tools to apply color adjustments for a cohesive aesthetic.
- **Transitions and Effects**: Options to smooth out cuts between scenes or add visual flair.
- **Text and Graphics**: To include captions, titles, or graphics overlay.
- **Music and Audio Editing**: Integrating background music, adjusting audio levels, and adding voiceovers.

3. Popular Mobile Video Editing Apps

3.1 iMovie (iOS)

iMovie is Apple's free video editing software available for iOS devices, providing a user-friendly experience suitable for beginners and casual editors.

Source: https://www.apple.com/in/imovie/

- **Features**
  - **Clip Trimming and Cutting**: Intuitive options to split and cut clips for easy editing.
  - **Pre-designed Themes and Transitions**: Includes pre-built themes, transitions, and music that add a professional touch with minimal effort.
  - **Trailer Templates**: Users can create movie-like trailers using built-in templates, an engaging feature for storytelling.
- **Example**
  - A small business owner creating a product demo could record footage using their iPhone and use iMovie to cut unwanted parts, apply a consistent color filter, add transition effects, and finish with a title sequence and background music.

- **Advantages**
  - Pre-installed on iOS devices, eliminating setup time.
  - User-friendly interface suitable for beginners.
- **Limitations**
  - Limited advanced features such as multi-track editing and detailed audio editing.

3.2 Kinemaster (iOS and Android)

Kinemaster is a robust app available for both Android and iOS, catering to more advanced users with a feature set approaching desktop software capabilities.

Source: https://kinemaster.com/en

- **Features**
  - **Multi-Track Editing**: Enables the layering of multiple video, image, and audio tracks.
  - **Advanced Effects and Transitions**: Provides chroma key for green screen effects, and a variety of transition styles.
  - **In-App Store**: Offers additional effects, stickers, fonts, and royalty-free music for a more customized experience.
  - **Precise Audio Controls**: Includes tools for adjusting audio levels, adding background music, and implementing voiceovers.
- **Example**
  - A content creator could use Kinemaster to produce a travel vlog, layering background music, adding captions, and using the chroma key feature to insert special effects. This app also allows for detailed control over transitions, giving the video a professional finish.
- **Advantages**
  - Advanced multi-layering options, ideal for more complex projects.
  - Available on both major mobile platforms, allowing cross-platform compatibility.

- **Limitations**
  - Some premium features require a subscription.
  - May be resource-intensive, requiring a newer device for optimal performance.

3.3 Adobe Premiere Rush (iOS and Android)

Adobe Premiere Rush is a mobile version of Adobe Premiere Pro, designed to integrate easily with Adobe's suite of products.

Source: https://www.adobe.com/in/products/premiere-rush.html

- **Features**
  - **Cross-Device Compatibility**: Users can edit on a mobile device and continue editing on a desktop.
  - **Drag-and-Drop Editing**: Intuitive drag-and-drop interface for arranging clips.
  - **Auto-Ducking for Audio**: Automatically adjusts background audio when dialogue is detected, which is ideal for vlogs and interviews.
  - **Cloud Storage**: Synchronizes projects across devices with Adobe's cloud.
- **Example**
  - A YouTube creator could film, edit, and publish short videos directly from their phone using Adobe Premiere Rush's tools, such as adding animated titles and adjusting audio levels. Later, they can open the same project in Adobe Premiere Pro on their desktop for further refinement.
- **Advantages**
  - Seamless integration with Adobe Creative Cloud for easy workflow across devices.

  - Equipped with essential video editing tools suitable for quick content creation.
- **Limitations**
  - Limited features compared to Adobe Premiere Pro.
  - Subscription required for full features and cloud storage.

3.4 InShot (iOS and Android)

InShot is a versatile video editing app that caters to social media creators with its straightforward editing tools and emphasis on platform-specific features.

Source: https://inshot.com/

- **Features**
  - **Aspect Ratio Presets**: Pre-configured aspect ratios for social media platforms such as Instagram, YouTube, and TikTok.
  - **Basic Editing Tools**: Includes trimming, splitting, merging, and duplicating clips.
  - **Text, Stickers, and Filters**: Extensive options for adding customizable text, emoji stickers, and a variety of filters.
  - **Speed Controls**: Allows users to speed up or slow down video playback for creative effects.
- **Example**
  - An Instagram influencer could use InShot to edit a quick story or reel, adding captions, adjusting the aspect ratio, and applying effects to match their personal style.
- **Advantages**
  - Quick, easy-to-use, and optimized for social media.
  - Comprehensive tools for basic editing and branding.

- **Limitations**
  - Limited to basic editing features, not suitable for extensive projects.
  - Watermark on videos in the free version.

4. Key Editing Techniques for Mobile Video Production

4.1 Trimming and Cutting

These are foundational skills in video editing. By cutting and trimming, users can eliminate unnecessary sections and focus on meaningful content. For example, removing dead space at the beginning and end of a video clip can significantly improve pacing and keep the audience engaged.

4.2 Color Correction and Filters

Using color correction tools can give a consistent aesthetic to video projects, making them more visually appealing. Apps like Adobe Premiere Rush and Kinemaster offer fine-tuning options such as brightness, contrast, saturation, and highlights, while simpler apps like InShot provide filters for a quick visual upgrade.

4.3 Adding Music and Voiceovers

Integrating background music or voiceovers can enhance storytelling and provide emotional cues. Kinemaster and iMovie allow for easy integration of audio tracks, with controls to adjust volume, balance, and fade.

4.4 Text and Graphics

Text overlays, such as titles and captions, provide context, especially for silent or tutorial videos. InShot and Adobe Premiere Rush make it easy to add text and graphic elements, which are especially useful for instructional videos or social media content.

4.5 Transitions and Effects

Transitions, such as fades and wipes, are useful for scene changes or introducing new topics. Overuse of transitions, however, can distract from the video's message. Apps like iMovie offer simple transitions, while Kinemaster provides more advanced options.

5. Considerations for Mobile Video Editing

5.1 Storage and File Management

High-resolution videos can take up significant storage space on a smartphone. Cloud-based solutions like Adobe Premiere Rush's Creative Cloud can help users manage storage by saving projects online.

5.2 Resolution and Export Settings

Exporting at the highest possible resolution ensures quality, especially for videos destined for larger screens or high-definition platforms. Most apps allow users to choose export settings based on platform requirements, ensuring compatibility with YouTube, Instagram, TikTok, and more.

5.3 Optimizing for Social Media Platforms

Different social media platforms have specific video requirements, from aspect ratios to time limits. Apps like InShot and Adobe Premiere Rush help users tailor their videos to meet these specifications, increasing the video's performance on each platform.

Conclusion

Mobile video editing apps have empowered users with tools that were once accessible only through desktop software. Apps like iMovie, Kinemaster, Adobe Premiere Rush, and InShot make it possible to produce high-quality, professional videos directly on a smartphone. While mobile apps may lack some of the advanced features of desktop software, their accessibility, affordability, and ease of use make them invaluable for on-the-go editing and instant publishing.

5

Synchronization of Audio in Video Editing Smart Phones (Mobile)

Introduction

In the era of digital media consumption, smartphones have emerged as powerful tools for video creation and editing making it accessible to everyone, from casual users to professional creators. With advancements in mobile technology, users can capture high-quality video and audio and edit them seamlessly within mobile applications. However, one of the significant challenges in video production is achieving proper synchronization between audio and video tracks. This chapter explores the principles of audio synchronization, the importance of this process in video editing, and practical techniques for achieving perfect alignment using smartphone applications.

Understanding Audio Synchronization in Mobile Video Editing

Audio synchronization, or "lip-syncing," is particularly vital in videos where dialogue, music, or sound effects are synchronized with visual cues. Accurate synchronization can help convey emotions, emphasize actions, and create an immersive experience. On smartphones, achieving perfect synchronization requires both intuitive tools and practical techniques to account for various factors, such as audio drift, frame rates, and video quality.

Smartphone applications for video editing typically support audio synchronization through visual audio waveform displays, alignment tools, and timing adjustments. These features empower users to manually or automatically adjust and align audio with visual footage.

Importance of Audio Synchronization

1. **Clarity**: Proper synchronization ensures that dialogue, sound effects, and background music align with the actions on screen. This clarity is essential for narrative coherence, especially in interviews, dialogues, and musical performances.
2. **Professionalism**: Well-synced audio contributes to the overall professionalism of the video. It reflects the creator's attention to detail and enhances the viewer's experience.

3. **Emotional Impact**: Synchronizing audio with video elements can amplify emotional responses. For example, the timing of sound effects with specific visual cues can evoke excitement, tension, or joy.

Common Challenges in Audio Synchronization

1. **Latency**: Latency occurs when there is a delay in the audio signal processing, leading to misalignment between sound and video.
2. **Multiple Audio Sources**: When using external microphones or multiple audio tracks, syncing can become complicated. Ensuring all sources align perfectly with the video requires careful attention.
3. **Editing Workflow**: The editing process can introduce synchronization issues, especially if audio is recorded separately from the video.

Essential Factors Affecting Audio Synchronization

Several technical aspects can impact audio synchronization on mobile devices:

1. **Frame Rate (Frames Per Second, FPS)**: Frame rates vary (e.g., 24, 30, 60 FPS) and directly impact how audio syncs with video. Inconsistent frame rates or mismatches between video and audio frame rates can lead to drifting or misalignment issues.
2. **Audio Sampling Rate**: Audio files generally have sampling rates of 44.1 kHz or 48 kHz. A mismatch in sampling rates between the recorded audio and video files can cause synchronization delays.
3. **Latency in Recording Devices**: When recording directly through a smartphone, especially with external microphones, audio delay or lag may occur due to latency in audio processing. Applications that offer latency compensation can help mitigate this issue.
4. **Compression and File Compatibility**: File types and compression formats affect synchronization. Highly compressed formats may sometimes cause sync issues, as they are more prone to audio drift during playback on smartphones.

Techniques for Synchronizing Audio in Mobile Video Editing

Mobile applications have made it easier for users to edit videos and synchronize audio effectively. Here are some techniques and best practices for achieving proper synchronization using smartphones:

1. Use of Visual Cues

One effective method for synchronization is to use visual cues. For example, during the recording, a visible clap or hand signal can serve as a point of reference. This technique is particularly useful when audio is recorded separately.

- **Implementation**: Record a visual cue (like clapping hands) in the video at the same moment the audio is captured. When editing, align the audio waveforms with the visual cue on the video timeline.

2. Waveform Matching

Many mobile editing applications display audio waveforms, allowing users to visually match peaks and troughs of sound waves with video events.

- **Implementation**: Zoom into the timeline of the video editing app and look for distinct audio waveforms that correspond to significant actions in the video. Align the audio tracks based on these visual cues for precise synchronization.

3. Frame-by-Frame Adjustment

For more complex edits where audio must sync with specific frames, frame-by-frame adjustments may be necessary.

- **Implementation**: In your editing app, use the frame-by-frame navigation feature to move the playhead. Fine-tune the audio track by nudging it left or right until it aligns perfectly with the visual events.

4. Use of Sync Tools in Editing Apps

Many modern video editing applications come equipped with synchronization tools that simplify the process. These tools automatically align audio tracks with video, making the process more efficient.

- **Examples of Mobile Apps with Sync Tools**
  - **KineMaster**: Offers a multi-layer timeline for audio and video, allowing users to drag and drop audio clips easily and align them with video events.
  - **Adobe Premiere Rush**: Features automatic audio sync for multi-camera setups, streamlining the process of aligning audio with video.
  - **iMovie**: Provides simple audio editing tools, allowing users to trim, adjust, and synchronize audio with video clips easily.

5. Editing Audio Directly on the Timeline

Most video editing applications enable users to edit audio tracks directly on the timeline. This can include adjusting volume levels, adding effects, and trimming unnecessary sections.

- **Implementation**: Once the audio is aligned, adjust the volume levels to ensure that dialogue is clear and background music complements rather than overwhelms the primary audio.

Key Mobile Applications for Audio Synchronization in Video Editing

Several video editing apps provide features for effective audio synchronization on smartphones. Below are some of the popular mobile applications:

1. KineMaster

- **Overview**: KineMaster is a versatile video editing app that supports multiple video and audio tracks, with precise controls for audio synchronization.
- **Sync Features**: It offers a "pre-trim" feature where users can adjust audio timing before adding it to the timeline. KineMaster also provides audio filters and effects for refining synced audio quality.
- **Benefits**: The app's multi-track capabilities make it suitable for complex audio syncing tasks, such as dialogue alignment and background score timing.

2. Adobe Premiere Rush

- **Overview**: Adobe Premiere Rush is a mobile and desktop app designed to simplify video editing workflows, including audio syncing.
- **Sync Features**: Premiere Rush allows for precise audio track positioning and uses a waveform view to help users visually align audio with video. It also supports importing high-quality audio formats to prevent compression-related sync issues.
- **Benefits**: Its compatibility with Adobe Premiere Pro enables users to sync audio and refine edits on a professional editing platform if needed.

3. InShot

- **Overview**: InShot is a user-friendly video editing app focused on short-form content, popular among social media users.
- **Sync Features**: InShot provides basic audio timing adjustments and visual waveform tools for aligning background music and effects with visual cues.
- **Benefits**: While limited in professional features, InShot is highly effective for quick edits and simple audio synchronization tasks for social media videos.

4. FilmoraGo

- **Overview**: FilmoraGo by Wondershare is an intuitive mobile video editor with multiple audio and video tracks, suitable for precise synchronization tasks.

- **Sync Features**: FilmoraGo allows users to cut, trim, and position audio tracks in sync with video content. It includes real-time previews, making it easier to spot sync issues during editing.
- **Benefits**: The app's timeline interface supports frame-by-frame audio adjustments, helping users align audio with visual actions seamlessly.

Best Practices for Audio Synchronization

1. **Plan Ahead**: When shooting, plan how audio will be recorded and ensure that any necessary equipment is in place. If using external microphones, test the audio input to avoid latency issues.
2. **Keep it Simple**: Limit the number of audio sources if possible. This reduces complexity during the editing phase and minimizes the chances of synchronization issues.
3. **Test and Review**: After editing, play the video back multiple times to ensure that the audio is perfectly synchronized. Look out for any discrepancies and make necessary adjustments.
4. **Utilize Headphones**: While editing, use headphones to better hear audio discrepancies and ensure clarity in sound levels.
5. **Export and Review**: Once the editing process is complete, export a draft of the video to review on different devices. This allows for identifying any sync issues that may not be apparent during editing.

Steps for Effective Audio Synchronization on Mobile Apps

The following steps can help ensure accurate audio synchronization during mobile video editing:

Step 1: Prepare Your Audio and Video Files

- Before importing audio into the editing app, ensure that the audio format and sampling rate match the requirements of the video.
- Convert audio files to compatible formats (e.g., WAV or MP3) with a standard sampling rate (48 kHz) for smoother synchronization.

Step 2: Import Files to the Timeline

- Import your video and audio files to separate tracks on the editing timeline.
- Zoom in on the timeline to visualize audio waveforms and pinpoint cues for precise alignment.

Step 3: Adjust Sync Using Visual Cues and Audio Waveforms

- Align audio and video by matching visual events (e.g., lip movements, hand claps) with corresponding audio cues in the waveform.

- Most editing apps display audio waveforms, which makes it easier to match audio events with visual actions.

Step 4: Test and Refine Synchronization

- Play back the edited video to check for sync errors, especially in scenes with dialogue or action.
- If any drifting occurs, adjust audio segments slightly until synchronization is perfect. Apps like KineMaster and Premiere Rush offer frame-by-frame adjustments to correct minor sync issues.

Troubleshooting Common Audio Sync Issues on Mobile

1. **Audio Drift**: Over longer videos, audio may gradually drift from the video. To solve this, cut the audio into smaller segments and align each segment with its corresponding video portion.
2. **Latency Delays**: When using external microphones, latency may affect synchronization. Using apps with latency compensation or pre-recording audio and video separately can help address this.
3. **Export and Compatibility Issues**: Sync issues may appear after exporting due to compression. Export videos in high-quality formats and test the final video on various platforms to ensure consistent sync.

Example Scenarios in Mobile Audio Synchronization

Example 1: Vlogging

- A travel vlogger records scenes using an external microphone connected to their phone. After importing video and audio to Premiere Rush, they notice slight delays in audio. Using Premiere Rush's waveform view, they align the audio with visual cues and fine-tune the sync with frame adjustments.

Example 2: Social Media Content Creation

- A content creator uses InShot to add a music track to a dance video for Instagram. They manually position the music to match the beat with their dance movements, trimming the track in sync with specific visual cues.

Example 3: Short Film Production

- An independent filmmaker uses KineMaster to produce a short film. They record audio separately using a high-quality device, then import both the video and audio files into the app, aligning dialogue and sound effects with the video. KineMaster's multi-track feature allows them to adjust sync precisely for each scene.

Mobile video editing has expanded the creative possibilities for both amateur and professional creators. However, achieving high-quality audio synchronization on a smartphone requires understanding both the app's tools and the techniques involved in aligning sound with video. From vlogging to short films, the accuracy of audio synchronization impacts the storytelling power of videos. By mastering synchronization techniques in mobile apps like KineMaster, Adobe Premiere Rush, and FilmoraGo, users can produce videos with cohesive audio-visual harmony, enhancing the overall quality and impact of their work.

6

Exporting
Final Video Output in Mobile App

Exporting a video is the final, crucial step in the video editing process. Whether you're producing content for social media, YouTube, or a personal project, exporting optimally ensures that the video plays seamlessly across different devices and platforms. This chapter will guide you through the export options available in popular mobile video editing apps, describe file formats and compression settings, and discuss ways to retain quality while optimizing file size.

Understanding Export Basics

When exporting video in a mobile editing app, several key factors affect the final output quality and compatibility. These include:

- **Resolution**: The clarity and detail of your video. Common resolutions include 720p (HD), 1080p (Full HD), and 4K (Ultra HD).
- **Frame Rate**: Number of frames per second (fps), affecting the smoothness of the video. Typical frame rates are 24, 30, and 60 fps.
- **Compression**: Reduces file size but can impact quality. Apps often offer adjustable compression settings.
- **Format**: Determines compatibility with devices and platforms. Standard formats include MP4, MOV, and AVI.

Example: Choosing the Right Export Settings

Suppose you've created a 2-minute travel video for Instagram using your mobile editing app. Choosing 1080p resolution at 30 fps with high compression in MP4 format will yield a high-quality video without an excessively large file size, suitable for social media.

Export Options in Popular Mobile Apps

Let's explore some widely used mobile video editing apps and the export settings they offer.

a) Adobe Premiere Rush

Adobe Premiere Rush, designed for quick and high-quality edits, allows users to export in several formats and resolutions.

- **Settings**: Premiere Rush offers exporting in 720p, 1080p, and 4K, with options for frame rates from 24 to 60 fps.
- **Example**: After editing, you can select "Share" in the menu, choose "Export" and opt for social platform-specific settings (like Instagram, YouTube). Adobe provides preset configurations that help you meet each platform's requirements effortlessly.

b) KineMaster

KineMaster is favored for its robust tools and versatile export options.

- **Settings**: Offers exporting in resolutions up to 4K with a maximum frame rate of 60 fps. It also provides adjustable bitrate and quality options.
- **Example**: For a short vlog, export in 1080p at 30 fps. This setting ensures a balance between quality and file size, essential for easy sharing on platforms like Facebook or Twitter.

c) InShot

InShot is beginner-friendly with a variety of export options.

- **Settings**: Allows resolutions up to 4K and adjustable bitrate, with export quality settings from "High" to "Low."
- **Example**: For a simple project like a tutorial video, export in 1080p at 24 fps in "High Quality" mode. This setup ensures clarity for instructional content without creating a massive file.

Best Practices for Exporting Video on Mobile

To ensure your video exports smoothly and is ready for viewing, follow these practices:

- **Test Export Settings**: Run a few test exports in different formats and resolutions to see which settings best balance quality and file size.
- **Check Platform Requirements**: For uploading to social media or other sites, check their recommended resolution, format, and bitrate to ensure compatibility.
- **Use Wi-Fi for Large Exports**: When exporting large videos, use a stable Wi-Fi connection to avoid interruptions during the process.

Example Scenario

Imagine you're exporting a 10-minute tutorial for YouTube. You choose 4K resolution at 24 fps in MP4 format with medium compression. You check YouTube's recommended settings and confirm they match. Once exported, upload directly to YouTube from the app or save to your device for later sharing.

Troubleshooting Common Export Issues

a) File Too Large

- **Solution**: Reduce resolution or increase compression. For instance, switching from 4K to 1080p can drastically reduce file size without sacrificing noticeable quality.

b) Slow Export Times

- **Solution**: Close background apps and ensure adequate device storage and battery. Heavy processing demands can slow down the export process.

c) Poor Quality on Playback

- **Solution**: Adjust bitrate and check playback device compatibility. Exporting in MP4 format often ensures compatibility across devices.

7

Creating Video Using Images and Text

Creating videos with images and text is a versatile way to convey information visually and is popular in many forms of media, from social media posts to professional presentations. This chapter covers the basics of making videos using images and text, discusses design techniques for a compelling visual narrative, and provides examples using popular tools.

Introduction to Image and Text-Based Videos

Image and text-based videos are commonly used for tutorials, presentations, promotional content, and storytelling. These videos rely on sequences of images combined with text overlays or captions to convey messages without needing video footage. This approach is ideal for visually-driven stories or information-focused videos.

Key Benefits

- **Simple to Produce**: Requires no video footage, relying instead on images and text, which can be arranged on mobile devices or desktop software.
- **Effective Storytelling**: Allows for clear and concise communication, ideal for instructional videos, presentations, or photo stories.
- **Flexible**: Can be used across platforms, from social media to formal presentations, providing versatility in application.

Steps to Create a Video Using Images and Text

Step 1: Plan the Video Storyboard

Begin by planning the structure of the video. Outline key points, visuals, and text to guide the viewer through your message. Consider the sequence in which the images will appear, any text overlays, and timing.

Example

For a "How-to" tutorial on gardening, plan out each step in sequence:

1. Introduction slide with title text.
2. Step 1: Choosing seeds (image of seeds with brief text).

3. Step 2: Planting (image of soil and tools).
4. Step 3: Watering and care (image of plants being watered).
5. Conclusion slide.

Step 2: Gather and Arrange Images

Gather high-quality images that match your storyboard. Images should be high resolution to maintain clarity. Arrange the images in the sequence planned in the storyboard.

Step 3: Add Text Overlays

Text is a crucial element in image-based videos as it guides the viewer's understanding of each image. Use text to highlight important details, emphasize key points, and provide context.

Tips for Effective Text Use

- **Keep It Brief**: Limit text to concise phrases to maintain readability.
- **Choose Contrasting Colors**: Use text colors that stand out against the background to ensure readability.
- **Placement**: Position text in areas that do not obstruct the main image elements.

Step 4: Use Transitions for Smooth Flow

Transitions between images can make the video more visually appealing. Simple fades, slides, or zoom-ins keep the viewer engaged and help the video flow smoothly.

Step 5: Add Background Music or Voiceover

Adding audio can enhance the viewer's experience. Background music adds mood, while a voiceover can narrate the content for further explanation.

Step 6: Adjust Timing and Finalize

Each slide's duration should align with the text's readability and the complexity of the image. Shorter slides can be used for simple images with brief text, while complex images may need more time.

Examples Using Popular Tools

There are numerous tools available for creating videos using images and text. Here are examples using three popular options: Canva, Adobe Spark, and iMovie.

Example 1: Canva

Canva's video editor is beginner-friendly and includes templates for text overlays, transitions, and music.

1. **Create a New Project**: Select the video format, choose a blank template, or start with pre-designed options.
2. **Upload Images**: Drag images into the timeline in the desired order.
3. **Add Text**: Use the text tool to overlay titles, subtitles, or descriptions.
4. **Transitions**: Apply transitions between each slide to make the video flow.
5. **Export**: Preview the video, adjust as needed, and export in your preferred format.

Source: wiki image

Example 2: Adobe Spark

Adobe Spark is designed for creating short, visually-appealing videos. It has built-in templates and allows easy text integration.

1. **Select a Template**: Choose a template that aligns with your content theme.
2. **Insert Images**: Place images into slides, and Adobe Spark automatically adjusts them.
3. **Add Text**: Add captions or explanations using the text tool.
4. **Add Music**: Use Adobe Spark's free music options or upload your own.
5. **Export**: Preview and export the video for download or direct upload to social media.

Source: https://adobe.fandom.com/wiki/Adobe_Spark

Example 3: iMovie (for iOS)

iMovie offers basic editing tools that are perfect for creating image and text videos on mobile.

1. **Create a New Project**: Open iMovie and select "Movie" to start a new project.
2. **Import Images**: Add images to the timeline in the preferred sequence.
3. **Text Overlays**: Choose the "Titles" option to add text overlays to each image.
4. **Transitions**: Select transition styles to enhance flow between slides.
5. **Export**: Save the video in various resolutions depending on where it will be shared.

Tips for Creating Engaging Image and Text Videos

a) Use High-Quality Images

The quality of images directly affects the final video. Blurry or pixelated images can detract from the message, so always use high-resolution images.

b) Keep Text Concise

Viewers often skim through videos, so keep text simple and impactful. Avoid overcrowding slides with too much text, which can overwhelm the viewer.

c) Choose the Right Music

Background music should complement the tone of the video. For an instructional video, light, upbeat music works well, while storytelling videos might benefit from more atmospheric tracks.

d) Maintain Consistent Style

Consistent font, color scheme, and layout provide a professional look to your video. Many apps offer theme options that maintain uniformity across slides.

8

Designing of Video Titles Using Adobe Photoshop (Text)

Video titles play a critical role in grabbing viewers' attention and establishing the tone and theme of the video. Adobe Photoshop, known for its robust text and graphic design capabilities, is an ideal tool for creating custom, professional video titles. In this chapter, we will explore the basics of title design, step-by-step instructions for using Photoshop to design titles, and tips to make text visually appealing and engaging for video content.

Introduction to Video Title Design

Video titles are the first visual cue that sets the audience's expectations. An effective title is visually appealing, represents the video's theme, and provides essential information, like the video's name or segment titles. Adobe Photoshop allows designers to combine typography, graphics, and special effects, creating dynamic title sequences that draw viewers in.

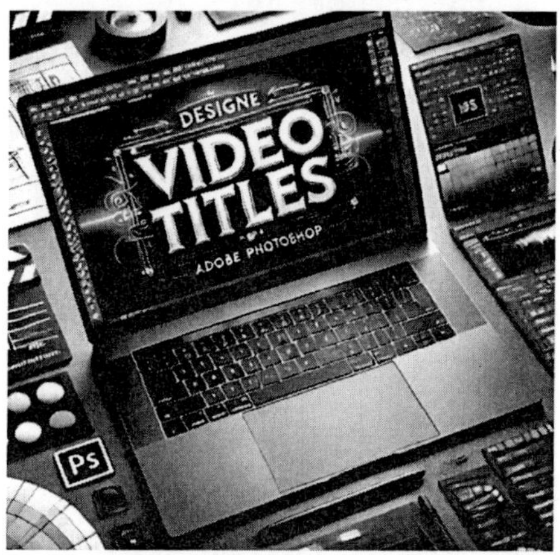

Key Elements of Effective Title Design

- **Font Choice**: Selecting a font that suits the video's theme is crucial. Serif fonts often suggest a formal tone, while sans-serif fonts are more modern and clean.

- **Color Scheme**: Use colors that contrast well with the background and remain readable on different screens.
- **Placement and Alignment**: Aligning text thoughtfully helps guide the viewer's eye and maintains visual balance.
- **Text Effects**: Effects like shadows, outlines, and 3D text can add depth and help the title stand out.

Steps to Design Video Titles Using Adobe Photoshop

Step 1: Set Up a New Project

1. Open Adobe Photoshop and create a new project. Set the canvas dimensions to match your video resolution, commonly 1920x1080 pixels for HD video.
2. Set the background color to transparent or choose a color based on the video's theme.

Step 2: Add and Customize Text

1. Select the **Text Tool (T)** from the toolbar and click on the canvas where you want to add text.
2. Type the title text, and choose a font style that aligns with the video's tone. Photoshop offers an extensive list of fonts, or you can import custom fonts if needed.
3. Adjust the font size, weight, and spacing to ensure readability.

Step 3: Apply Text Effects

1. With the text layer selected, go to **Layer > Layer Style** to add effects like drop shadow, stroke, and bevel.
2. **Drop Shadow**: Creates depth by casting a shadow behind the text, making it pop on light or busy backgrounds.
3. **Outer Glow**: Adds a glow around the text, useful for darker backgrounds.
4. **Gradient Overlay**: Allows blending colors within the text for a more dynamic, eye-catching title.

Step 4: Add Graphics or Icons (Optional)

Adding small graphics or icons can further enhance the title design. For instance, adding icons related to a theme (e.g., a camera icon for a photography title) makes the title more visually engaging.

1. Import graphics or shapes by going to **File > Place Embedded**.
2. Position and resize them in relation to the text.

Step 5: Save and Export

1. Once satisfied with the title design, export it in a video-compatible format. Choose **File > Export > Save for Web (Legacy)** and select PNG format with transparency preserved for seamless overlay on video.
2. Alternatively, save the file in PSD format to retain editable layers if future adjustments are needed.

Tips for Creating Engaging Titles

1. **Use Contrasting Colors**: Ensure that the title stands out against the video's background.
2. **Minimal Text**: Keep titles concise. Longer titles can appear cluttered and may lose the viewer's attention.
3. **Animate Titles**: Adobe After Effects or Premiere Pro can animate Photoshop titles for dynamic effects.
4. **Test Readability**: View titles on different screens to confirm legibility across devices.

9

Importing Text Titles from Photoshop in Video Editing

Text titles play an essential role in creating compelling video content by providing context, enhancing storytelling, and engaging viewers. While most video editing software offers basic title features, Adobe Photoshop allows video creators to design intricate, customized titles with enhanced visual effects. Once created, these titles can be imported into video editing software, where they are integrated into the final video. This chapter explores the steps for importing text titles from Photoshop into video editing platforms, discussing techniques, examples, and best practices to ensure high-quality results.

Understanding Text Titles and Their Role in Video

Text titles serve various purposes in video production, such as conveying information (e.g., names, locations, subtitles), adding emphasis, and supporting branding. High-quality titles create a polished look and help maintain viewer interest. Photoshop provides a comprehensive toolkit for crafting these titles, including tools for color grading, typography, textures, and other effects, which video editing software often lacks.

Steps for Creating and Exporting Titles in Adobe Photoshop

Step 1: Setting Up the Canvas

- Start by opening Photoshop and creating a new project with the same dimensions as your video (e.g., 1920x1080 pixels for full HD). Setting up the canvas with the correct dimensions ensures the title appears well-proportioned in the video.

Step 2: Designing the Title

- Use Photoshop's text tool to type the title and explore various font styles. Adjust typography settings like kerning, leading, and size for readability.
- Add stylistic elements such as shadows, strokes, and textures. Titles can benefit from effects like gradients, outlines, or drop shadows to make them stand out.

Step 3: Exporting the Title

- After completing the design, make the background layer transparent by hiding it, ensuring that only the text is visible when overlaid in the video.
- Export the title as a PNG file to maintain transparency or as a PSD file if the video editing software supports Photoshop files (e.g., Adobe Premiere Pro).

Importing Titles into Video Editing Software

Each video editing software has unique steps for importing text files, but the general process is similar across platforms.

Example: Importing into Adobe Premiere Pro

- Open your video project in Premiere Pro.
- Go to File > Import and select the PNG or PSD file created in Photoshop.
- Drag the imported title onto the timeline, positioning it on a separate layer above the video clip where it should appear.

Example: Importing into DaVinci Resolve

- Open DaVinci Resolve, navigate to the Media tab, and import the title PNG file.
- Place the file on the timeline in the Edit workspace, adjusting its position and duration as needed.

Tips for Adjusting Imported Titles

1. **Resizing and Positioning**
   - Use the video editor's transform tools to resize or reposition the title to fit the frame perfectly.
2. **Adding Animation**
   - Most editors, like Premiere Pro or DaVinci Resolve, allow you to animate imported titles. You can create fade-in/fade-out effects, slide animations, or scale transformations.
3. **Combining with Backgrounds or Overlays**
   - Some titles work well with additional background or overlay effects. For example, semi-transparent shapes behind the title text can improve readability over busy video scenes.
4. **Color Correction**
   - To ensure the title colors match the video, consider minor color grading adjustments directly in the video editing software.

Examples of Effective Title Design

- **Intro Titles:** Short text titles at the start of the video, often including brand logos and introductory text.
- **Lower Thirds:** Titles used during interviews or vlogs to identify speakers.
- **End Credits:** Often created as simple, scrolling text but can be designed in Photoshop for a unique style.

Challenges and Best Practices

1. **Maintaining Quality During Import**
   - Use high-resolution PNG files to avoid quality degradation.
2. **Avoiding Complex Effects**
   - While Photoshop offers many effects, not all transfer well in video editing. Avoid effects like gradients or glows that may not render consistently.
3. **File Compatibility**
   - Ensure the format (PNG or PSD) is compatible with the editing software. Some platforms may not fully support PSD layers or effects.

10

Voice Recording Techniques Using Smart Phones

Introduction

In the digital age, smartphones have evolved into powerful tools capable of handling various tasks, including high-quality audio recording. This capability has revolutionized voice recording for multiple applications, from podcasts to interviews, video voiceovers, and field recordings. This chapter explores techniques for optimizing voice recording using smartphones, along with an overview of mobile apps and equipment to enhance audio quality. By the end of this chapter, you will have a comprehensive understanding of effective voice recording practices using your smartphone, regardless of your skill level.

Importance of High-Quality Voice Recording

Clear and professional voice recording is essential in creating engaging audio content. Poor audio quality can deter audiences, even when paired with high-quality video or visuals. Smartphone voice recording, however, has unique challenges: background noise, limited microphone quality, and recording consistency. Understanding techniques and choosing suitable equipment and apps can substantially improve audio quality and make a lasting impression.

Basic Techniques for Voice Recording with Smartphones

Choose a Quiet Recording Environment

Background noise is a common issue in smartphone recording. Reducing unwanted sounds can significantly enhance recording quality. Ideally, select a quiet, enclosed space, and turn off any sources of noise (fans, air conditioners, etc.). For better sound insulation, soft furnishings and rugs help absorb unwanted ambient noise.

Proper Phone Positioning and Microphone Direction

The built-in microphone on smartphones is generally located near the bottom of the device. For optimal recording, position the phone at a slight angle, about 6-12 inches from your mouth, to avoid excessive noise while ensuring clarity.

Speaking directly into the microphone can improve the sound's richness, while using a pop filter (or even a cloth) helps mitigate popping sounds caused by sharp consonants.

Avoid Unnecessary Movements

Movement can create unwanted rustling sounds and volume inconsistencies in recordings. Placing the phone on a stable surface or using a tripod helps maintain steady audio quality. For interviews or podcasts, keeping the phone as stationary as possible ensures consistent audio.

Advanced Techniques for Enhancing Audio Quality

Using External Microphones

Smartphones support a variety of external microphones that improve audio quality beyond the built-in options. Some popular choices include:

- **Lavalier (lapel) microphones:** Small and clip onto clothing, ideal for interviews or vlogging.
- **Shotgun microphones:** Capture directional sound, reducing surrounding noise.
- **USB condenser microphones:** Plug into newer smartphones and offer enhanced audio detail.

Each type has its benefits depending on the recording scenario. Lavalier mics work well for voice clarity, while shotgun mics are ideal for isolated, directional recording.

Utilize Noise Reduction Features

Many mobile apps offer built-in noise reduction features that help filter background noise during recording or in post-production. Apps like Dolby On and AudioLab provide noise reduction filters that can significantly enhance the clarity of recorded audio.

Recording in High-Quality Formats

Most smartphones record audio in standard formats like MP3 or M4A, but for professional-grade projects, recording in uncompressed formats like WAV or AIFF is preferred. High-quality formats capture more detail and allow for better editing, making them ideal for podcasts, interviews, and video voiceovers.

Recommended Smartphone Apps for Voice Recording

Various mobile apps provide extensive features for enhanced audio quality:

- **Dolby On:** Offers noise reduction, de-essing, and EQ adjustments, suitable for vocal recording.
- **Anchor by Spotify:** Aimed at podcast creators, it features simple editing and the ability to add background music.
- **AudioLab:** Provides advanced editing features and format conversions.
- **Ferrite Recording Studio:** Designed for journalists and podcasters, with multi-track recording capabilities.

These apps provide various controls to enhance audio quality and enable basic editing on the go.

Editing Recorded Audio on Smartphones

Once recorded, post-production can refine the audio further. Most mobile apps offer editing features such as:

- **Noise reduction filters:** Remove background sounds.
- **Equalizer (EQ) adjustments:** Balance frequencies to improve vocal clarity.
- **Cut and trim:** Remove unwanted segments to polish the final audio.

Applying these edits allows for a professional finish, even when recorded on a smartphone.

Tips for Effective Voice Recording with Smartphones

Warm-Up and Practice

Before recording, warm up your voice with vocal exercises, especially for lengthy recordings or formal presentations. Practicing and adjusting volume levels beforehand helps achieve consistent sound levels.

Avoid Overlapping Voices

When recording multiple speakers, encourage participants to avoid talking over each other, as this can create audio overlaps that are challenging to edit.

Use Headphones for Real-Time Monitoring

Using headphones during recording allows for real-time monitoring and adjustment, which can help catch issues with audio quality and noise early.

Smartphones have democratized voice recording, making it accessible to anyone with a mobile device. While smartphones have limitations, using the right techniques, apps, and external accessories can significantly enhance the quality of your recordings. Practicing and refining these techniques will ensure that your voice recordings meet professional standards.

11

Installing of Video and Audio Software into Desktop Computer

Introduction

As multimedia creation continues to expand, having robust video and audio editing software on desktop computers has become essential for professionals, hobbyists, and students. Installation of these software applications is the foundational step, enabling users to leverage the advanced capabilities of desktop-based editing. This chapter will explore the process of installing video and audio software, considering system requirements, compatibility, and essential installation steps. It will also provide troubleshooting advice and recommendations for some commonly used software options in the video and audio editing industry.

Understanding System Requirements

Before installing any video or audio editing software, it is critical to understand the software's minimum and recommended system requirements. This can ensure smooth operation and prevent potential issues such as crashes, lagging, and poor performance.

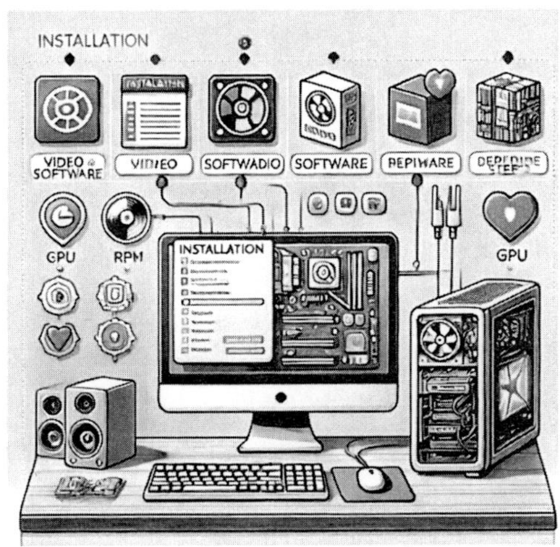

Key Hardware and Software Requirements

- **Processor (CPU):** A multi-core processor, such as Intel i5 or AMD Ryzen 5 and above, is often recommended. Certain programs, especially those for professional editing, may benefit from Intel i7/i9 or AMD Ryzen 7/9 processors.
- **Memory (RAM):** At least 8GB of RAM is generally recommended for basic editing, while 16GB or more is ideal for more complex projects.
- **Graphics Processing Unit (GPU):** A dedicated graphics card, such as NVIDIA GTX or AMD Radeon, will significantly enhance video rendering speeds and allow for smoother previews.
- **Storage:** Software installation requires a considerable amount of hard disk space, with additional space necessary for project files. SSDs are preferred over HDDs for faster data access and improved performance.
- **Operating System Compatibility:** Verify that the software is compatible with your operating system, be it Windows, macOS, or Linux. Some software applications are platform-specific, while others offer cross-platform support.

Choosing Video and Audio Software

Selecting the right software depends on several factors, including budget, functionality, and intended use. Many high-quality programs are available, both as open-source and premium options.

Popular Video Editing Software

- **Adobe Premiere Pro:** A leading choice for professional video editors, known for its extensive editing tools and seamless integration with other Adobe software.
- **DaVinci Resolve:** This powerful software offers advanced color correction and video editing, with a free version that includes numerous professional features.
- **Final Cut Pro:** Available only on macOS, this software is popular among professional editors for its performance and intuitive interface.

Popular Audio Editing Software

- **Audacity:** A free, open-source audio editor with comprehensive editing features. Ideal for recording, cutting, and refining audio tracks.
- **Adobe Audition:** Part of Adobe's Creative Cloud suite, Audition is known for its multitrack editing capabilities and integration with Adobe Premiere Pro.

- **FL Studio:** Primarily used for music production, FL Studio offers audio editing and mixing capabilities as well.

Installation Steps for Video and Audio Software

Here's a general step-by-step guide on how to install video and audio editing software on your desktop.

Step 1: Downloading the Software

1. Visit the official website of the software you want to install.
2. Check for the latest version compatible with your OS and download the installer. Ensure you download from the official site to avoid malware or altered versions.
3. Once downloaded, locate the installer file in your "Downloads" folder or the location you specified.

Step 2: Running the Installer

1. Double-click the installer file. The installation wizard should open automatically.
2. Follow the prompts to begin the installation. You may be asked to agree to the software license agreement.
3. Select the installation location on your hard drive, which is usually the default "Program Files" directory for Windows or "Applications" folder for macOS.
4. Some installers provide options for custom installation, allowing users to select specific features or plug-ins.

Step 3: Configuring Settings Post-Installation

1. **Initial Setup:** Upon first opening, many software programs will run an initial setup to detect hardware capabilities and configure settings automatically.
2. **Setting Preferences:** Go to "Preferences" to configure settings such as language, workspace layout, and shortcut keys.
3. **Audio/Video Input and Output Devices:** Set up any external devices, such as microphones or cameras, by navigating to the audio/video input settings.

Troubleshooting Installation Issues

Even with a straightforward installation process, you may encounter issues. Here are some common problems and solutions:

- **Software Crashing Upon Startup:** Verify that your system meets the minimum requirements. Update your graphics drivers and ensure your operating system is up to date.
- **Compatibility Issues with OS Updates:** If the software is not compatible with the latest OS update, check if there is a patch available on the software's official website.
- **Permissions Issues on macOS:** macOS may restrict installations from unrecognized developers. To bypass this, go to "System Preferences" > "Security & Privacy" > "General" and select "Open Anyway" after attempting installation.
- **Error Messages:** Search for specific error codes in the software's support forum or knowledge base for potential fixes.

Additional Resources

To fully utilize video and audio editing software, familiarize yourself with support resources offered by developers. Most software providers have online manuals, video tutorials, and community forums.

1. **Official Documentation and Tutorials:** Websites like Adobe, DaVinci Resolve, and Audacity offer comprehensive documentation and step-by-step video tutorials.
2. **Online Learning Platforms:** Websites such as Udemy, LinkedIn Learning, and Skillshare offer courses on video and audio editing basics, as well as specific tutorials for software like Premiere Pro and DaVinci Resolve.
3. **YouTube Channels:** Channels like "Adobe Creative Cloud" and "Film Riot" feature tutorials and tips that are helpful for beginners and advanced users alike.

12

Installing of Audio Recording Equipment with Sound Card

Installing audio recording equipment on a desktop computer requires a sound card setup and proper equipment configuration for effective audio capture and playback. This chapter covers the essentials of setting up an audio recording environment, focusing on sound card installation, microphone setup, and configuration of other essential audio equipment for achieving professional-quality recordings.

Overview of Sound Cards in Audio Recording

Sound cards, also known as audio interfaces, play a crucial role in audio recording by converting analog audio signals into digital data for the computer and vice versa. They also manage audio input and output and allow the computer to communicate effectively with microphones, speakers, and other audio equipment.

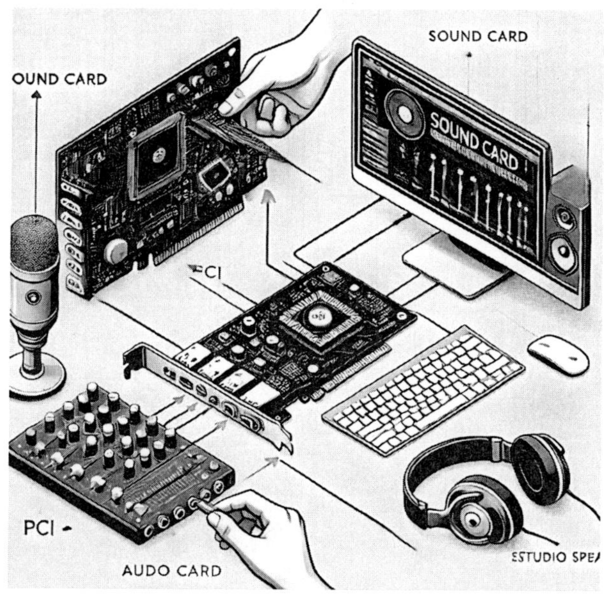

There are two main types of sound cards used for recording:

- **Internal Sound Cards**: Installed directly on the computer's motherboard, typically through a PCI or PCIe slot.
- **External Sound Cards (Audio Interfaces)**: Connected via USB, Thunderbolt, or FireWire and are generally favored for higher-quality audio recording and portable use.

Example Equipment: Popular audio interfaces include the Focusrite Scarlett series, the Behringer UMC series, and the PreSonus AudioBox.

Selecting the Right Sound Card for Your Needs

When selecting a sound card for audio recording, consider the following factors:

- **Bit Depth and Sample Rate**: A higher bit depth (e.g., 24-bit) and sample rate (e.g., 48 kHz or higher) allow for clearer audio.
- **Input and Output Ports**: Consider the number of audio inputs/outputs you'll need. For example, a two-input interface is enough for solo recording, but multi-channel options are available for more extensive setups.
- **Compatibility**: Make sure the sound card is compatible with your operating system and audio recording software.
- **Latency**: Lower latency is crucial for real-time audio monitoring without delays.

Steps for Installing an Internal Sound Card

Requirements: You will need a sound card, screwdriver, and an anti-static wrist strap.

1. **Turn off and Unplug the Computer**: Ensure the computer is completely powered off and unplugged to avoid electrical hazards.
2. **Open the Computer Case**: Use a screwdriver to remove the side panel of the computer.
3. **Locate the PCI Slot**: Identify an available PCI or PCIe slot on the motherboard, depending on your sound card type.
4. **Install the Sound Card**: Insert the sound card into the PCI slot, applying even pressure until it is securely seated.
5. **Secure the Card**: Use a screw to attach the sound card to the case, securing it firmly.
6. **Close the Case**: Reattach the side panel and plug the computer back in.

7. **Install Drivers**: Turn on the computer and install the necessary drivers for the sound card. Most drivers are available on the manufacturer's website.

Setting Up an External Audio Interface

1. **Connect the Audio Interface**: Use a USB, FireWire, or Thunderbolt cable to connect the audio interface to your computer.
2. **Install Drivers and Software**: Install any required drivers, which can usually be downloaded from the manufacturer's website.
3. **Configure the Interface**: Open your computer's audio settings or Digital Audio Workstation (DAW) and select the audio interface as the primary sound device.

Example: If using Audacity, go to Edit > Preferences, select Audio Devices, and choose the connected audio interface.

Configuring Microphones and Other Audio Equipment

1. **Microphone Setup**: Connect a microphone to the audio interface. For condenser mics, enable phantom power (+48V) on the audio interface.
2. **Headphone and Speaker Setup**: Connect headphones to the audio interface for monitoring and speakers if available.
3. **Test Audio Levels**: Use your DAW or sound card software to test and adjust audio levels. Make sure that the sound is clear and without distortion.

Example Setup: In Adobe Audition, go to Audio Hardware under Preferences, select the audio interface, and set input/output levels for optimal sound.

Configuring Software for Audio Recording

Setting up audio recording software requires attention to several key settings to ensure quality and efficiency. Here's a step-by-step guide to optimizing software for audio recording:

1. **Selecting Audio Devices**
   - In your digital audio workstation (DAW) or recording software, navigate to the audio settings menu.
   - Select the desired sound card or audio interface as the input and output device. This ensures your software captures sound from your preferred hardware.
   - For systems that support it, ASIO drivers are recommended as they provide low-latency performance ideal for recording.

2. **Setting Sample Rate and Bit Depth**
   - For most professional audio recordings, a sample rate of 44.1 kHz or 48 kHz and a bit depth of 24 bits are standard. These settings balance quality and file size.
   - Higher sample rates (e.g., 96 kHz) may be used for high-fidelity projects, though they increase processing load.
3. **Configuring Buffer Size**
   - Buffer size affects latency (the delay between input and playback). A smaller buffer reduces latency, which is ideal for live monitoring but may strain CPU resources.
   - For recording, use a low buffer size to avoid delay. During mixing and editing, you can increase it to reduce CPU load.
4. **Input and Output Routing**
   - Configure your DAW to route audio inputs and outputs effectively. For instance, assign microphones to specific input channels and monitor output through headphones or speakers.
   - Some DAWs allow custom routing setups, useful for complex recording needs.
5. **Adjusting Gain Levels**
   - Check input gain levels to prevent distortion. Set microphone input levels so peaks remain below 0 dB to avoid clipping.
   - Use software gain adjustments sparingly to avoid adding noise; instead, adjust hardware gain settings on your audio interface if possible.
6. **Monitoring and Latency Management**
   - Enable direct monitoring if your audio interface supports it, allowing you to hear the recording without software-induced latency.
   - If direct monitoring is unavailable, ensure your DAW latency is minimized by using optimized settings.
7. **Configuring Tracks and Channels**
   - Set up individual tracks for each audio source, such as vocals, instruments, or background sounds.
   - Name tracks for easy identification and apply panning to separate sounds spatially in the final mix.

8. **Setting Up Effects and Plugins**
   - Most audio recording software supports plugins (e.g., VST, AU) for added effects like reverb, equalization, and compression.
   - Add these plugins post-recording to keep initial recordings clean. Non-destructive editing ensures the original audio remains intact.
9. **Saving and File Management**
   - Save your project regularly in a dedicated folder with organized subfolders for audio files, plugins, and temporary files.
   - Consider backing up projects to avoid data loss.
10. **Testing**
    - Record a short test track and play it back to confirm all settings are configured correctly.
    - Listen for audio quality, any unexpected noise, and latency issues. Fine-tune settings as needed.

For optimal recording, configure your DAW (e.g., Audacity, Adobe Audition or Pro Tools) to recognize the sound card or audio interface. Adjust the buffer size and latency settings within the software to ensure smooth playback and avoid delays.

- **Setting Sample Rate and Bit Depth**: For example, set the sample rate to 44.1 kHz or higher for music and audio quality settings to 24-bit.
- **Adjusting Latency**: In the DAW, find the latency settings to adjust for real-time audio monitoring without lag.

Troubleshooting Common Issues

Installing audio recording equipment with a sound card can come with several common issues, many of which have straightforward troubleshooting steps:

1. **Driver Issues**: The most frequent problem arises from outdated or missing drivers for the sound card. Ensure the latest drivers from the manufacturer's website are installed, and if the issue persists, try reinstalling the drivers or checking for operating system compatibility.
2. **Hardware Compatibility**: Some sound cards may not be compatible with certain motherboards or operating systems. Before installation, confirm that the sound card matches the hardware requirements of your system. BIOS settings may also need to be adjusted to recognize the sound card.
3. **Improper Configuration**: Incorrect input/output settings in the operating system or digital audio workstation (DAW) software can

prevent audio signals from reaching the recording equipment. In both the system's audio settings and within your DAW, ensure that the sound card is selected as the primary input/output device.

4. **Latency Issues**: Latency can cause delays in audio playback and recording. To resolve this, adjust the buffer size in your DAW or use ASIO drivers, which offer low-latency performance tailored for audio recording.
5. **Electrical Interference and Noise**: Buzzing or hissing noises may result from electrical interference, poor grounding, or low-quality cables. Use shielded cables, connect to a grounded power source, and ensure that the sound card is properly seated in the PCI slot.
6. **Firmware Conflicts**: Sometimes, updates to the sound card's firmware may conflict with system software. Check for any firmware updates and be cautious about compatibility before installing these updates to prevent issues with connectivity or stability.

By addressing these common issues, most problems with installing audio recording equipment alongside a sound card can be quickly resolved, allowing for high-quality audio recording and playback.

13

Exporting Final Video in Different Formats and Sizes

Introduction

Exporting video is the final stage in the production pipeline and serves as the bridge between the editing environment and the audience. To ensure quality and compatibility across various devices, formats, and platforms, it is essential to understand the nuances of video export settings, including video resolution, bitrate, codec, and file size. This chapter provides a detailed overview of exporting video files in different formats and sizes, covering a variety of settings used in professional video editing software.

Understanding Video Formats and Codecs

- **Definition of Video Formats**: Video formats define how data is stored within a file. The most commonly used formats are MP4, MOV, AVI and MKV.
- **Role of Codecs**: Codecs, short for *coder-decoder*, compress and decompress video files to save storage space. Common codecs include H.264, H.265 (HEVC), and VP9.
- **Choosing the Right Format**: Different platforms may require different formats, e.g., MP4 for YouTube and social media, MOV for Apple-based environments, and AVI for Windows environments.
- **Compatibility Across Devices**: Devices and platforms vary in compatibility. Selecting the right codec and format ensures smooth playback on target devices without sacrificing quality.

Resolution and Aspect Ratios for Exporting

- **Common Resolutions**
  - **SD (Standard Definition)**: 720x480 pixels, often used for DVDs.
  - **HD (High Definition)**: 1280x720 and 1920x1080, widely used for online video content.
  - **4K (Ultra HD)**: 3840x2160, used for high-end productions.

- **Aspect Ratios**: Important to ensure that video appears correctly on different screens. The standard aspect ratios include 16:9 for widescreen, 4:3 for traditional TV, and 9:16 for vertical mobile videos.
- **Considerations for Exporting in Multiple Resolutions**: Exporting in different resolutions allows the same content to be viewed on both high-end and low-end devices.

Bitrate and Compression

- **Understanding Bitrate**: Bitrate is the amount of data processed per second of video. It directly affects both quality and file size.
- **Types of Bitrates**
    - **CBR (Constant Bitrate)**: Suitable for streaming as it maintains a consistent rate.
    - **VBR (Variable Bitrate)**: Adjusts bitrate based on scene complexity, yielding higher quality at smaller sizes.
- **Balancing Compression and Quality**: High bitrate yields better quality but larger file size. Compression techniques (e.g., H.264, H.265) can help balance quality and file size for efficient storage and streaming.

Exporting for Specific Platforms

Exporting videos tailored to specific platforms ensures compatibility, optimized playback, and improved viewer experience. Each platform often has its preferred formats, resolutions, and aspect ratios, so adjusting these settings beforehand can enhance the performance and visual appeal of the final video.

Social Media (YouTube, Instagram, Facebook): These platforms generally recommend H.264 codec with MP4 format at 1080p or higher for the best playback.

1. **YouTube**: For YouTube, export in H.264 format (MP4), ideally at a 16:9 aspect ratio for widescreen playback. Recommended resolutions are 1080p (1920x1080) or 4K (3840x2160) for HD and Ultra HD viewing. YouTube supports bitrates between 8-12 Mbps for HD, and higher for 4K. Audio should be in AAC format with a bitrate of 128 kbps or higher.
2. **Instagram**: Instagram has varied requirements for posts, stories, and IGTV. For standard posts, square (1:1) or vertical (4:5) formats work best. Stories are vertical at 9:16 (1080x1920). Use H.264 codec with a maximum bitrate of 5 Mbps to ensure smooth mobile streaming. Keep videos under 1 minute for posts and up to 15 seconds for stories to fit platform limits.

3. **Facebook**: Export for Facebook in MP4 format with a 16:9 aspect ratio (or 4:5 for mobile). HD resolution (1080p) is recommended with a bitrate of 6-8 Mbps. Audio should be in AAC format, and videos should ideally be under 240 minutes long.
4. **Twitter**: Twitter supports H.264 MP4 files, best at 1280x720 resolution with a maximum bitrate of 5 Mbps. Videos should ideally be in landscape (16:9) but can also work in square (1:1) for mobile viewing. Twitter limits videos to 2 minutes and 20 seconds in duration.
5. **TikTok**: TikTok videos are vertical (9:16) and should be exported at 1080x1920 resolution, H.264 format. Bitrates between 2-5 Mbps work well for TikTok's platform, with a maximum duration of 3 minutes. Lower resolutions (720p) are also acceptable for faster upload and smoother playback.

Web: For web, formats like MP4 (H.264 or VP9 codec) are preferred due to their balance of quality and load time.

Broadcast: Requires high-quality, high-bitrate formats like ProRes or DNxHD to ensure minimal loss in broadcast environments.

Adhering to these platform-specific settings during export ensures videos look professional and load efficiently across devices, maximizing engagement and playback quality.

Export Settings in Popular Software

- **Adobe Premiere Pro**: Offers export options that range from H.264 for social media to high-quality ProRes files for broadcast. Key settings include resolution, frame rate, bitrate, and audio quality.
- **Final Cut Pro**: Provides versatile export options, including custom settings for Apple devices, high-end broadcasting, and social media optimization.
- **DaVinci Resolve**: Known for its advanced export panel, DaVinci Resolve supports formats from H.264 to uncompressed RAW.

Step-by-Step Exporting Process

Exporting a video ensures that all edits, effects, and settings are processed into a final file format, ready for viewing or sharing. Here's a simple step-by-step guide to exporting a video:

1. **Finalize the Timeline**: Ensure all edits are complete, transitions are smooth, and no elements need adjustment. Double-check for any missing media or placeholder files.

2. **Choose Export Settings**: Go to your software's export or render option. Select output settings based on where the video will be viewed:
   - **Resolution** (e.g., 1080p, 4K) depends on the desired video quality and platform requirements.
   - **Aspect Ratio** (e.g., 16:9, 1:1) should match the viewing platform (widescreen, square for social media, etc.).
   - **Format** (e.g., MP4, MOV) should align with platform compatibility and preferred quality settings.
3. **Select the Codec**: Choose a video codec like H.264 for high quality with manageable file sizes or ProRes for editing quality. Codecs affect compression and playback quality.
4. **Set Bitrate**: Define the bitrate (often in kbps or Mbps) for the video and audio. Higher bitrates yield higher quality but increase file size. Lower bitrates are preferable for faster uploads and streaming.
5. **Audio Export Settings**: Configure audio settings by selecting the format (e.g., AAC for smaller sizes or WAV for uncompressed quality), sampling rate, and bitrate to match your output goals.
6. **Preview the Export**: Most software offers a preview option to ensure colors, framing, and quality appear as expected.
7. **Export and Save**: Click export and wait for the render to complete. This process varies in duration based on video length, resolution, and computer speed. Save the output file to an organized location for easy access.
8. **Review and Test**: Once exported, play the video on different devices or platforms to confirm playback quality. Check for any color shifts, audio issues, or resolution drops.

Easy to Remember

- **Step 1**: Set the sequence or timeline resolution based on your final output requirements.
- **Step 2**: Choose the export format and codec, with H.264 being a popular option for online and MP4 format.
- **Step 3**: Configure bitrate settings (VBR or CBR) and specify target and maximum bitrates.
- **Step 4**: Set audio quality, generally around 320 kbps for high-quality audio.
- **Step 5**: Check file size estimates, quality, and compatibility settings before finalizing export.

Advanced Export Settings: HDR, 360° Video, and VR

- **HDR (High Dynamic Range)**: Allows for a wider range of colors and contrast. Supported by modern platforms such as YouTube. HDR enhances video quality by increasing the range of color and contrast. When exporting in HDR, editors select settings compatible with HDR-capable devices and platforms, ensuring rich, lifelike visuals. For HDR output, formats like HLG (Hybrid Log-Gamma) and PQ (Perceptual Quantizer) are commonly used, and bit depth is typically set to 10-bit or higher.
- **360° Video**: For an immersive, spherical video experience, 360° video must be exported in an *equirectangular* format, which allows the video to wrap around a virtual environment. High resolution (4K or more) is recommended to maintain detail as viewers pan across the video. 360° videos are popular on platforms like YouTube and Facebook, where viewers can engage interactively.
- **VR (Virtual Reality)**: VR video export requires high resolution and frame rates (ideally 60 fps or more) to reduce motion sickness and provide a smooth experience. VR files are often exported in stereoscopic (for 3D depth) and monoscopic formats, with spatial audio for realistic soundscapes. This setting enables VR videos to be compatible with VR headsets like Oculus or HTC Vive.

File Size Considerations and Storage Management

- **Reducing File Sizes Without Compromising Quality**: Techniques include reducing bitrate, using compression-friendly codecs, and lowering resolution for smaller screens.
- **Storage and Backup Options**: Large files require robust storage solutions such as cloud storage (Google Drive, Dropbox), or external drives with high-speed read/write capabilities.

Exporting the final video is a critical step in the video production workflow that requires careful attention to technical settings. By understanding various formats, resolutions, and codecs, editors can optimize the viewing experience for different platforms. This knowledge allows editors to enhance the quality of their work, ensuring compatibility across devices, efficient storage, and high-quality playback. Proper export settings elevate the final output and contribute significantly to the viewing experience.

14

Voice Synchronization to Video Using Advanced Software

Introduction

Voice synchronization, or "lip-syncing," involves matching audio tracks to video so that spoken words align precisely with the movements in the video. This is crucial in various production contexts, such as film, animation, video dubbing, and post-production editing, where accurate synchronization enhances the viewing experience by making interactions feel natural and seamless.

Advanced video editing and audio software, like Adobe Premiere Pro, DaVinci Resolve, and Avid Media Composer, provide powerful tools to achieve this with precision. These tools include automated time-stretching, audio wave alignment, and frame-by-frame adjustments, allowing editors to visually align audio waveforms with video frames for meticulous timing. Additionally, AI-driven tools, such as those found in software like Adobe Audition or Descript, can further streamline synchronization by automatically detecting and adjusting timing mismatches, saving time while improving accuracy.

Proper synchronization in video production ensures that the audio and visual elements are cohesively blended, enhancing storytelling and engaging the audience. For professionals, this attention to detail in syncing voice to video is essential for achieving high-quality, polished final products across various media.

This chapter provides an in-depth look at the tools and techniques for voice synchronization using advanced software, including Adobe Premiere Pro, DaVinci Resolve, Avid Media Composer, and Adobe Audition. We will discuss step-by-step processes, professional tips, and troubleshooting methods that can enhance the syncing process and ensure professional-quality results.

Understanding Voice Synchronization Basics

Voice synchronization requires meticulous alignment between the audio and visual components of a project. Key points include:

- **Timing Adjustments:** Achieving the perfect timing of dialogue to match the lip movements and actions in the video.
- **Waveform Analysis:** Visualizing audio waveforms to help align specific parts of dialogue with exact video frames.
- **Frame-by-Frame Syncing:** Using frame-accurate adjustments for minor timing discrepancies, a feature provided by most advanced software.

The importance of accurate voice synchronization extends beyond aesthetics; it is critical in creating immersive, coherent, and engaging media.

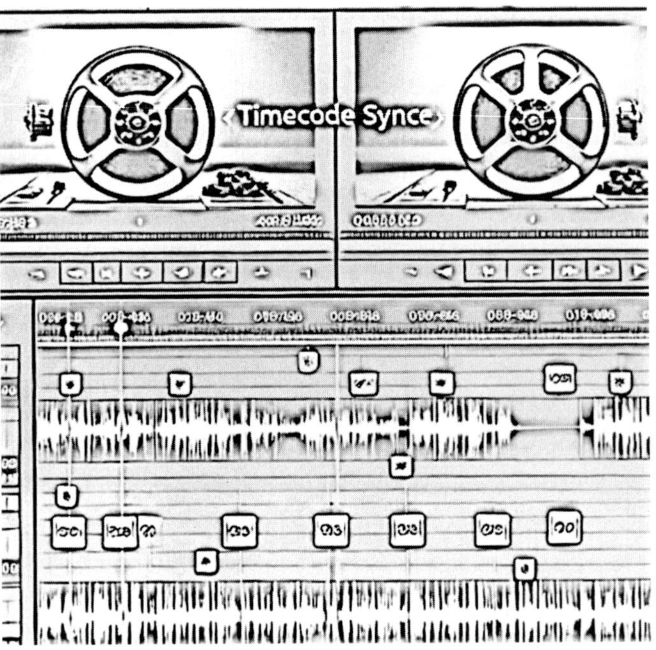

Principles of Voice Synchronization

- ***Matching Audio Peaks:*** Software can use audio waveforms to detect and align peaks in audio (e.g., dialog or specific sounds) with corresponding visual frames.
- ***Lip-Syncing:*** The importance of aligning spoken words with visible mouth movements in video. Challenges include managing differences in speech patterns, accents, and timing.
- ***Latency Compensation:*** Some devices introduce audio or video delay, which can throw off sync. Advanced software allows users to adjust and compensate for these delays to achieve frame-accurate synchronization.

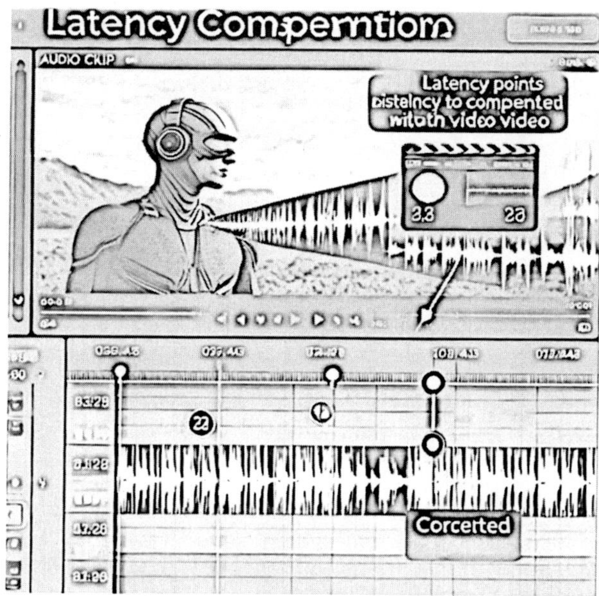

Voice Synchronization Techniques

- Timecode-Based Synchronization:
  - How it Works: A shared timecode between audio and video sources ensures precise alignment, often required for multi-language dubbing or high-stakes productions.
  - Software Support: Many advanced editing suites support timecode import and synchronization.
- Waveform-Based Synchronization:
  - How it Works: Audio waveforms are compared, and patterns in the voice track are matched to ensure sync. This technique is common in documentary and live event editing, where timecodes may be absent.
  - Example: Adobe Premiere's "Synchronize by Audio" feature enables voice and video tracks to be matched by analyzing waveform peaks.

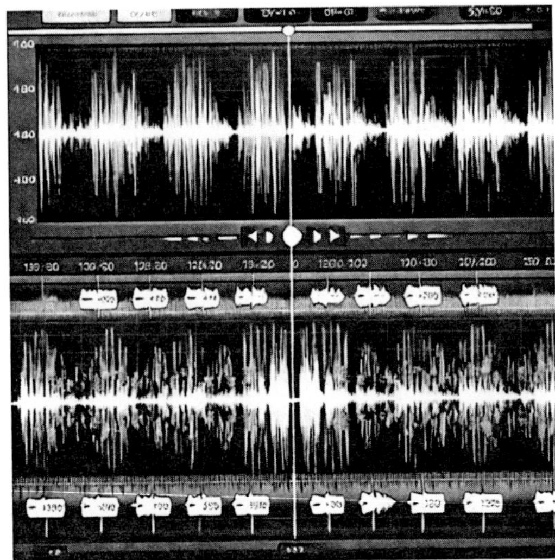

- Manual Synchronization with Markers:
  - How it Works: When automatic syncing fails, editors can use markers to match specific points manually in the audio and video.
  - Practical Use: Particularly useful for voice-overs or cases where sync drift occurs.

Voice Synchronization Workflow in Advanced Software

Software Options for Voice Synchronization

Modern video editing software provides a range of tools for fine-tuning audio synchronization. Key software options include:

- **Adobe Premiere Pro:** Premiere offers comprehensive syncing capabilities, including the ability to view audio waveforms alongside video frames and tools for manual adjustment of audio clips.
- **DaVinci Resolve:** Known for its robust editing and color grading features, DaVinci Resolve also includes powerful audio syncing options, like waveform syncing and timecode matching.
- **Avid Media Composer:** This software offers robust audio editing features suitable for large-scale productions. It allows editors to fine-tune audio timing with frame-by-frame precision.
- **Adobe Audition:** Often used for detailed audio editing, Audition can enhance synchronization in post-production, including pitch-shifting, stretching, and compression to fit dialogue with visuals.

Key Techniques for Voice Synchronization

There are several techniques professionals use to synchronize voice to video effectively.

Waveform Alignment

Waveform alignment is one of the most accurate ways to sync audio to video, especially when using dialogue. The process involves:

1. **Viewing the Waveform:** Audio editing software displays the waveform as a visual representation of the audio. This helps editors match peaks and valleys to visual cues on screen.
2. **Marking Key Frames:** Place markers at significant points, such as sentence starts or pronounced words, to use as anchor points when syncing the audio to the video.
3. **Adjusting Clip Timing:** Adjust audio clips frame-by-frame to ensure the dialogue aligns with the mouth movements in the video.

Using Timecode and Synchronization Tools

Advanced software like Avid Media Composer and DaVinci Resolve allows editors to use timecode and synchronization tools to automatically align audio and video tracks:

- **Timecode Matching:** When both audio and video have timecode embedded, the software can automatically align them.
- **Auto-Sync Tools:** Many software packages offer automated syncing options based on waveform and timecode, significantly reducing manual effort.

Frame-by-Frame Adjustments

For minor timing mismatches, frame-by-frame adjustments can help achieve perfect alignment. This technique is especially useful for fine-tuning lip-sync during dialogue sequences.

Speed and Pitch Adjustment

Some software offers tools to adjust the speed and pitch of audio slightly, helping synchronize the voice with the visual pacing. Adobe Audition, for instance, provides pitch-shifting and time-stretching tools that help fine-tune audio clips without compromising quality.

Step-by-Step Process

1. Importing Media: Import video and audio files into your editing software (e.g., Adobe Premiere, DaVinci Resolve, Final Cut Pro).

2. Initial Syncing: Use the software's synchronization feature (e.g., "Sync by Audio") to align audio and video tracks based on waveforms or timecodes.
3. Fine-Tuning: Adjust sync manually if needed, focusing on lip movement alignment with speech.
4. Latency Compensation: Apply any necessary delay adjustments in the software settings to correct sync drift.
5. Playback and Verification: Review the synced content to confirm alignment; use split-screen views if available.

Troubleshooting Common Synchronization Issues

1. **Audio Lag or Delay:** Audio may lag or lead slightly in the timeline, resulting in unsynchronized dialogue. Frame-by-frame adjustments can help remedy this issue.
2. **Background Noise Interference:** Excessive background noise can disrupt audio clarity. Using noise reduction tools in Adobe Audition or DaVinci Resolve can improve clarity.
3. **Pitch or Speed Distortion:** Adjusting pitch or speed to synchronize audio can occasionally lead to distortion. Ensure only minor adjustments are made to maintain natural-sounding dialogue.

Common Challenges in Voice Synchronization

- Audio Drift: Audio and video may go out of sync over time, especially with long recordings. Advanced software allows for segmenting or stretching clips to maintain sync.
- Frame Rate Inconsistencies: Different frame rates in audio and video can result in sync issues. Editors need to confirm consistent frame rates or use frame rate conversion tools.
- Lip Movement Mismatches: Non-native language dubbing often presents timing challenges, as the original speech patterns may differ from the dubbed audio. Software with real-time editing allows for frame-by-frame adjustments.

Advanced Software Features for Voice Synchronization

- Automatic Audio Syncing: Discussing Adobe Premiere's and Final Cut Pro's capabilities in matching audio and video based on waveform recognition.
- Multi-Layered Sync Views: DaVinci Resolve and other software offer split views, enabling side-by-side or stacked clips for precise adjustments.

- Frame-by-Frame Adjustment Tools: Many software options provide detailed timecode views, allowing for frame-by-frame control over audio positioning.
- AI-Assisted Syncing: Tools like Adobe's Sensei use AI to automatically detect and adjust voice sync, potentially learning speech patterns for more accurate alignment.

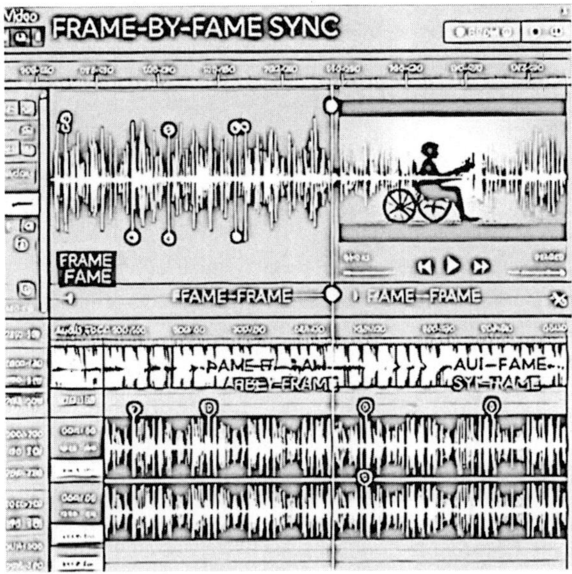

Case Studies and Practical Examples

India

1. **Bollywood Film Dubbing:** The Bollywood film industry, renowned for its dubbing needs across multiple languages, has adopted advanced voice synchronization software to streamline its post-production process. Bollywood films are often dubbed in regional languages and even adapted for international audiences, making synchronization vital. For instance, Yash Raj Films and Red Chillies Entertainment utilize software like **VoiceQ** and **Adobe Premiere Pro** with advanced audio alignment features. These tools help speed up the lip-syncing process, reducing the traditional manual time requirements and allowing quicker, cost-effective production across multiple languages.
2. **E-learning and EdTech Companies:** Companies such as **BYJU's** and **Vedantu** use synchronized voice-over techniques to localize educational videos for students across different regions. They employ AI-powered software to match regional language voice-overs with visuals, ensuring

accurate lip movements and seamless user experiences. **Synchro Arts' Revoice Pro** and **Descript** are commonly used for this purpose, helping to create interactive, language-specific learning materials.

3. **Indian News and Media Outlets:** News networks like **NDTV** and **Times Now** regularly synchronize interviews and video reports to regional dialects to increase accessibility. With AI synchronization tools like **Trint** and **Sonix**, editors can quickly match dubbed audio to the original video footage, improving the clarity and timeliness of broadcast content. This software speeds up translation time while maintaining high-quality audio-video alignment, a feature particularly useful in time-sensitive reporting.

Western World

1. **Hollywood Film Dubbing and Localization:** Hollywood relies heavily on voice synchronization for international film distribution. Studios like **Disney** and **Warner Bros.** have embraced AI-powered tools such as **Veed.io** and **Adobe Audition** to ensure precise voice synchronization across languages for their major film releases. For instance, Disney uses advanced synchronization software to ensure that animated characters' lips are synced perfectly with dubbed languages, enhancing audience immersion worldwide.
2. **Gaming Industry:** The gaming sector, particularly with titles developed by companies such as **Ubisoft** and **Electronic Arts (EA)**, uses advanced voice synchronization software for localizing in-game audio in multiple languages. Games like *Assassin's Creed* and *The Sims* are popular globally and require audio synchronization to match character movements with various language versions. **Wwise** and **iZotope RX** help streamline this complex task by ensuring audio matches visual actions with high precision, elevating the gaming experience for users worldwide.
3. **Streaming Platforms and Virtual Reality (VR):** Platforms such as **Netflix** and **Amazon Prime** use voice synchronization for dubbed content. AI-powered tools, including **Avid Pro Tools** and **DaVinci Resolve**, facilitate seamless synchronization for foreign films, particularly when matched with 4K and HDR video standards. In VR, where immersive experiences are paramount, synchronization tools help match spatial audio to 360° visuals. **Spatial audio plugins** in Avid Pro Tools, for example, ensure that the viewer's experience remains engaging and realistic.

4. **Documentary Productions:** Documentaries in the Western world often utilize synchronization tools to ensure that interviewees' audio matches visuals accurately. Productions like **BBC's Planet Earth** and **National Geographic's** documentaries use software such as **Nuendo** and **Descript** to handle complex audio needs, enabling creators to integrate synced voice-overs and translations efficiently. This process ensures that narrations align seamlessly with nature visuals or footage of interviewees, delivering a cohesive viewing experience.

These case studies underscore the broad application and growing demand for advanced voice synchronization software, especially as media becomes increasingly globalized. By applying these AI tools, industries from film to education achieve efficient, high-quality audio-video synchronization, meeting diverse viewer needs and expanding their content's accessibility across languages and cultures.

Future of Voice Synchronization Technology

- AI-Powered Synchronization: AI-powered synchronization has revolutionized audio-visual editing by automating the once painstakingly manual process of matching audio to video. By harnessing artificial intelligence algorithms, video editing software can now analyze and match lip movements, gestures, and even subtle facial expressions to the corresponding audio in real time. This technology improves efficiency, precision, and accessibility for content creators, streamlining workflows across fields like film, television, gaming, and online content production.

Some advanced video editing platforms now incorporate AI algorithms to speed up synchronization. Tools like Adobe Premiere Pro's Speech-to-Text, DaVinci Resolve's Voice Isolation, and platforms such as Descript use AI to analyze spoken words and automatically match them to audio waveforms, generating frame-accurate alignment without human intervention. This enables editors to fine-tune content faster and eliminates many typical synchronization challenges, such as delays, manual waveform matching, and frame-by-frame adjustments.

In addition to accuracy and time savings, AI-powered synchronization contributes to improved lip-syncing for dubbed content and voice-over adjustments, where language changes require precise synchronization. For creators working with large volumes of content, this technology simplifies voice and sound matching while ensuring high quality and consistency.

- Real-Time Sync for Streaming: As live streaming grows, real-time voice sync tools are becoming increasingly advanced, ensuring high-quality audio sync during broadcasts.

- Multilingual AI Dubbing: With the rise of AI dubbing, multilingual synchronization will be revolutionized, enabling real-time translation and dubbing synced accurately to video.

Professional Tips for Effective Synchronization

1. **Preview Regularly:** Constant previewing ensures the quality of the synchronization throughout the editing process.
2. **Use Headphones for Clarity:** High-quality headphones help detect subtle mismatches, especially with dialogue and ambient sounds.
3. **Engage in Precision Editing:** Small adjustments go a long way in achieving high-quality sync. Frame-by-frame editing is essential for any precise synchronization work.

Voice synchronization is a crucial skill for anyone involved in video production, especially as audiences expect high standards in both audio and visual elements. By mastering these tools and techniques, video editors can create polished, professional-quality projects that captivate audiences and effectively convey the intended narrative. Leveraging advanced software like Adobe Premiere Pro, DaVinci Resolve, and Adobe Audition, editors can achieve synchronization with precision, ultimately enhancing the quality and engagement of their video content.

15

Working with Video Libraries and Graphics Libraries

Introduction

Video and graphics libraries are essential tools for creating, managing, and enhancing multimedia projects, from professional filmmaking to web development. These libraries provide pre-built elements, including animations, images, icons, effects, and stock video footage, which can significantly reduce production time and elevate the quality of visual content. In this chapter, we will discuss the role of video and graphics libraries, popular library platforms, techniques for integrating these assets into projects, and essential considerations when working with them.

Understanding Video Libraries

Video libraries offer a vast repository of footage that creators can use to supplement or enhance their projects. These libraries are especially useful in situations where creating original footage is impractical, costly, or time-consuming.

Key Elements in Video Libraries

- **Stock Footage:** Pre-recorded video clips covering a wide range of themes, from nature and urban landscapes to action shots and abstract animations.
- **Background Footage:** Visuals intended to support on-screen text or act as a backdrop.
- **Transitions and Effects:** Clips or assets used to transition smoothly between scenes or add effects to existing footage.
- **Licensed Footage**: Footage that's licensed for use under specific terms, such as Creative Commons or commercial licenses.

Popular Video Library Platforms

- **Pexels and Unsplash**: Known for high-quality, free video clips suitable for commercial and personal projects.

- **Videvo and Pond5**: Offer a mix of free and premium stock footage with extensive libraries covering various categories.
- **Artgrid and Shutterstock**: Feature high-resolution, royalty-free video clips and stock footage ideal for commercial video production.

Understanding Graphics Libraries

A **graphics library** is a software toolkit or framework that provides tools for creating, manipulating, and rendering images, 2D or 3D graphics, and visual effects in applications. These libraries are often used in game development, data visualization, user interface design, and multimedia applications.

Graphics libraries offer a variety of visual assets, including icons, illustrations, templates, textures, and customizable graphics, enabling designers to quickly develop consistent and visually appealing content.

Key Elements in Graphics Libraries

- **Icons and Symbols:** Scalable vector graphics representing common symbols, used in applications or websites.
- **Illustrations and Templates:** Pre-designed illustrations or templates for projects such as social media posts, presentations, and infographics.
- **Textures and Patterns:** Graphics used to create depth and complexity in backgrounds or digital artworks.
- **Vectors and Customizable Shapes:** Editable graphics that can be scaled without loss of quality, commonly used in design software.

Popular Graphics Library Platforms

- **Flaticon and Icons8**: Icon libraries providing a range of customizable icons in various styles and formats.
- **Freepik and Canva**: Offer both free and premium illustrations, templates, and vectors.
- **Envato Elements and Adobe Stock**: Provide extensive libraries of graphics, including templates, photos, and illustrations for both personal and commercial use.

For 2D Graphics

1. **Pygame** (Python): Simple library for game development and 2D rendering.
2. **Matplotlib** (Python): Primarily for data visualization, but can be extended for 2D graphics.

3. **Processing** (Java): Designed for artists and visual designers for creative coding.
4. **Cairo** (C/C++): Vector graphics library with bindings for multiple languages.

For 3D Graphics

1. **OpenGL** (C, C++): A widely used standard for rendering 2D and 3D graphics.
2. **Vulkan** (C): A modern alternative to OpenGL with better performance for 3D rendering.
3. **DirectX** (C++, Windows): A Microsoft framework for rendering 3D graphics.
4. **Three.js** (JavaScript): Simplifies WebGL for creating 3D graphics in browsers.

For Cross-Platform Game and Multimedia Development

1. **Unity** (C#, C++): A game engine that supports 2D and 3D graphics.
2. **Unreal Engine** (C++): Known for high-quality rendering, widely used in gaming and simulation.
3. **Godot** (C++, GDScript): Open-source engine supporting 2D and 3D graphics.

For UI and Animation

1. **Qt Graphics View** (C++): Part of the Qt framework for building GUIs and 2D graphics.
2. **Skia** (C++): Used in Chrome, Android, and Flutter for fast and flexible graphics.
3. **D3.js** (JavaScript): For dynamic and interactive data visualizations.

Graphics Libraries for Specific Platforms

1. **Metal** (Swift, Objective-C, macOS/iOS): A low-level API by Apple for 3D rendering and computation.
2. **SFML** (C++): Simple and fast multimedia library for 2D graphics, sound, and input handling.

Other Notable Mentions

1. **Blender's Python API**: For creating and rendering 3D graphics programmatically.
2. **PixiJS** (JavaScript): For fast and advanced 2D WebGL-rendered graphics.

Integrating Video and Graphics Libraries into Projects

Integrating elements from these libraries into your project requires proper techniques to ensure cohesion and quality. This section covers a step-by-step approach for importing and using assets from video and graphics libraries effectively.

Step 1: Select the Appropriate Library

Choosing the right library depends on the project's requirements. For example, stock video footage for a travel blog might require vibrant nature shots, while a corporate project might need formal, clean backgrounds. When it comes to graphics, select styles that align with the branding or visual identity of your project.

Step 2: Importing and Organizing Assets

Import assets into your editing software, such as Adobe Premiere Pro, Final Cut Pro, or After Effects for video projects, or Photoshop and Illustrator for graphics projects. Create a dedicated folder structure within the software to keep assets organized by type (e.g., footage, icons, and transitions).

Step 3: Customizing and Editing Assets

Many video and graphics library assets can be customized to match the project's style. Adjust colors, apply filters, crop, or animate graphic elements. For example, in video editing software, stock footage may be layered with effects, while in design software, colors and sizes of icons or vectors can be modified.

Step 4: Ensure Quality and Cohesion

When incorporating multiple assets from different libraries, consistency is key. Ensure that color grading, resolution, and style are unified across the project. Apply standard filters or LUTs to video clips to maintain consistency, and use a color palette for graphic elements.

Best Practices for Using Video and Graphics Libraries

- **Verify License Restrictions:** Review usage licenses carefully, especially when using free assets, to ensure compliance with copyright and commercial usage guidelines.
- **Optimize File Size for Web and Mobile**: For digital and mobile-friendly projects, compress video and graphic files to ensure fast loading times without sacrificing quality.
- **Blend Multiple Assets Seamlessly:** Integrate assets creatively, blending stock elements with original content to maintain authenticity.
- **Maintain Quality Across Platforms**: High-definition (HD) or 4K formats are recommended for video elements to ensure quality on large screens.

Future of Video and Graphics Libraries

Video and graphics libraries are evolving to include advanced assets, such as 3D models, augmented reality (AR) visuals, and AI-generated content. Libraries are also beginning to offer more interactive assets, allowing for customization that can better fit unique project needs. As technology advances, creators can expect more adaptive and responsive tools that cater to VR, AR, and other immersive media.

Conclusion

Incorporating video and graphics libraries into creative workflows can enhance project quality and significantly reduce time spent on design and production. With the right tools and a well-organized approach, creators can leverage these resources to produce visually compelling, professional-grade content across different mediums and platforms.

16

Recording Techniques: Video and Audio Online and Offline Editing

Introduction

Video and audio editing encompass two primary modes: online and offline editing, each critical for producing high-quality, synchronized media content. Online editing refers to real-time editing with high-resolution files, typically as a final step in the post-production workflow. Offline editing involves working with lower-resolution copies or proxies of footage for easier manipulation and organization, saving time and computational resources. Both techniques hold unique advantages and can be integrated depending on project requirements, available resources, and editing goals.

Section 1: Understanding Offline Editing

1.1 Overview of Offline Editing

Offline editing is the preliminary editing stage where editors work with lower-resolution copies of the original footage to perform initial edits. This approach optimizes efficiency, as low-resolution files consume less storage and processing power. Common in projects with extensive footage, such as feature films, documentaries, and complex multimedia presentations, offline editing allows editors to focus on the structure, sequencing, and timing of content without concerns over resource demands.

1.2 Advantages of Offline Editing

- **Reduced Processing Power**: Editing with proxy or low-resolution files significantly reduces the strain on the computer, allowing faster rendering times and smoother timeline playback.
- **Efficiency in Long Projects**: With large-scale projects, offline editing is invaluable as it allows editing large files without burdening storage capacity or processing speed.
- **Enhanced Flexibility**: Editors can make comprehensive changes to the structure and pacing of the video without waiting for high-resolution playback.

1.3 Workflow for Offline Editing

1. **Ingesting and Converting Files**: Original high-resolution files are ingested into the editing software and converted into proxy files or low-resolution formats.
2. **Organizing Assets**: A structured asset organization simplifies the editing process and is essential for managing large projects.
3. **Editing in Proxy Mode**: Editors arrange the low-resolution footage to build the sequence, focusing on structure, cuts, and transitions.
4. **Relinking to High-Resolution Files**: Once the final sequence is ready, editors relink proxy files to the original high-resolution footage for detailed adjustments in the online editing stage.

1.4 Popular Offline Editing Software

- **Adobe Premiere Pro**: With proxy features, Adobe Premiere Pro allows editors to switch between high and low-resolution files seamlessly.
- **DaVinci Resolve**: Resolve's proxy workflow ensures efficient offline editing, especially for color grading.
- **Final Cut Pro**: Equipped with optimized media functions, Final Cut Pro is a strong choice for offline editing.

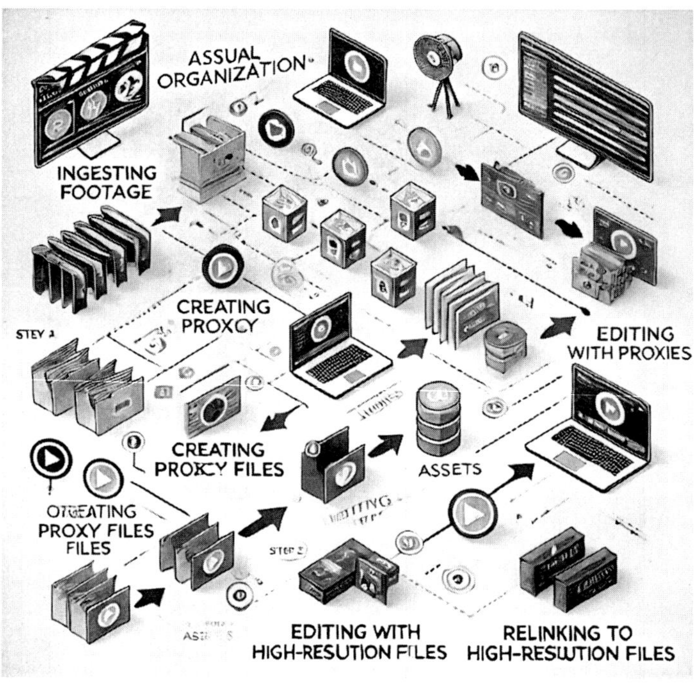

Section 2: Understanding Online Editing

2.1 Overview of Online Editing

Online editing is the final stage of editing, where high-resolution media files are used for color grading, sound synchronization, and final edits. This stage is critical to achieving professional quality and ensures all media elements are polished and synchronized. Online editing is labor-intensive but necessary to maintain the high quality of the final product.

Types of Online Editing

1. Document Editing

- **Platforms**: Google Docs, Microsoft Word Online, Dropbox Paper.
- **Features**:
  - Real-time collaboration.
  - Version history and tracking.
  - Auto-saving to cloud storage.
  - Compatibility with various file formats (e.g., DOCX, PDF).
- **Applications**
  - Academic writing.
  - Business documentation.
  - Collaborative content creation.

2. Video Editing

- **Platforms**: Canva Video Editor, Clipchamp, WeVideo, Adobe Premiere Rush.
- **Features**
  - Timeline-based editing.
  - Support for transitions, effects, and audio overlays.
  - Exporting in multiple resolutions.
  - Cloud storage for projects.
- **Applications**
  - Social media content creation.
  - Marketing videos.
  - Tutorials and educational content.

3. Image Editing

- **Platforms**: Canva, Pixlr, Photopea, Fotor.

- **Features**
  - Layer-based editing (Photopea).
  - Basic tools like cropping, resizing, and color adjustments.
  - Filters, overlays, and templates for quick edits.
  - Integration with cloud storage (Google Drive, Dropbox).
- **Applications**
  - Creating graphics for websites and social media.
  - Photo retouching.
  - Designing promotional material.

4. Code Editing

- **Platforms**: CodePen, JSFiddle, GitHub Codespaces, Replit.
- **Features**
  - Syntax highlighting for various programming languages.
  - Collaboration with multiple developers in real-time.
  - Integration with Git and version control systems.
  - Deployment and hosting options for web apps.
- **Applications**
  - Prototyping and testing code.
  - Collaborative coding projects.
  - Building and hosting small applications.

5. Audio Editing

- **Platforms**: TwistedWave, Soundtrap, BandLab.
- **Features**
  - Multitrack audio editing.
  - Built-in sound effects and loops.
  - Export in various formats (e.g., MP3, WAV).
  - Collaborative editing for podcasting or music production.
- **Applications**
  - Podcast editing.
  - Music production.
  - Voice-over enhancement.

2.2 Advantages of Online Editing

- **High-Quality Output**: Online editing enables precise adjustments to visual details, color grading, and effects at the highest resolution, resulting in a polished final product.
- **Accurate Synchronization**: Sound synchronization with high-resolution video files ensures professional results, essential for projects requiring high-quality audio-visual integration.
- **Final Adjustments and Effects**: Complex visual effects, transitions, and high-quality graphics are applied in this stage, finalizing the overall aesthetic of the video.

2.3 Workflow for Online Editing

1. **Importing High-Resolution Files**: Editors import the original, high-resolution files to replace proxy files or low-resolution footage.
2. **Finalizing Cuts and Transitions**: Any final changes to cuts or transitions are made with high-resolution footage to ensure precise results.
3. **Adding Color Grading and Effects**: High-quality color grading and visual effects enhance the video's appeal and match its thematic requirements.
4. **Sound Synchronization**: Audio is meticulously synchronized, with sound effects, music, and dialogues aligned for high clarity and impact.
5. **Final Export**: The completed video is exported in the required format, ready for distribution.

2.4 Popular Online Editing Software

- **Adobe Premiere Pro**: Premiere Pro's comprehensive tools for color grading, effects, and high-resolution exporting make it suitable for online editing.
- **DaVinci Resolve Studio**: Known for its advanced color grading capabilities, Resolve Studio is a top choice for finalizing high-quality video content.
- **Avid Media Composer**: This software is widely used in the film industry for online editing due to its precise editing tools and support for high-resolution media.

Section 3: Key Techniques in Video and Audio Editing

3.1 Video Editing Techniques

- **Cutting and Sequencing**: Editors arrange clips to tell a coherent story, paying attention to continuity, pacing, and rhythm.

- **Color Grading**: Adjusting colors enhances mood and atmosphere. With high-resolution footage in online editing, editors can apply professional-grade color correction.
- **Adding Visual Effects**: Effects like slow-motion, overlays, and green screen compositions are applied in the online editing phase for precision.

3.2 Audio Editing Techniques

- **Synchronization**: Aligning voiceovers, dialogues, and sound effects with video footage is essential for cohesive storytelling.
- **Sound Balancing and Equalization**: Adjusting audio levels and applying equalization ensure clarity across dialogue, background music, and sound effects.
- **Noise Reduction and Enhancement**: Removing background noise and enhancing audio quality is particularly important for online editing, where high-quality sound is expected.

Section 4: Best Practices in Combining Online and Offline Editing

- **Effective Asset Management**: Keeping files organized with clear naming conventions for both proxy and original files facilitates seamless transitioning between offline and online editing.
- **Maintaining Project Continuity**: Avoiding drastic changes between offline and online phases ensures consistency in cuts and transitions.
- **Backup and Version Control**: Storing versions of edits and regularly backing up files reduces the risk of data loss and allows for easy reversion to earlier edits.

Section 5: Challenges and Solutions in Online and Offline Editing

5.1 Common Challenges

- **System Overload**: Handling high-resolution files in online editing can strain computer resources.
- **Synchronization Issues**: Proper audio-video synchronization can be challenging, especially when working with multiple audio sources.
- **Storage Demands**: High-resolution video files require substantial storage capacity, making asset management crucial.

5.2 Solutions

- **Proxy Workflows**: Using proxies in offline editing reduces strain on resources, allowing complex projects to be edited on less powerful systems.

- **External Storage Solutions**: Investing in external storage drives or cloud storage eases storage constraints.
- **Automation Tools**: Some software includes tools for automated synchronization, which is particularly useful for interviews or multi-camera setups.

Conclusion

Mastering both online and offline editing techniques is essential for producing professional-quality videos and audio. Offline editing offers a flexible, resource-efficient way to structure a project, while online editing refines and enhances the final output. By understanding and effectively utilizing both modes, editors can create polished, high-quality media that meets the demands of modern viewers.

17

Recording Techniques: Video and Audio Linear & Non-Linear Editing

In the world of media production, recording techniques are the foundation upon which all post-production processes are built. Whether you're capturing video for a film, podcast, or YouTube video, understanding how video and audio are recorded and edited is crucial for achieving a polished and professional final product. In this chapter, we will explore the two main types of editing processes—linear and non-linear—and how they apply to both video and audio recording. By the end of this chapter, you'll have a thorough understanding of these editing methods and how to apply them effectively in your projects.

1. The Basics of Recording Techniques

Recording techniques refer to how video and audio are captured, whether it be on film, digital media, or any other format. For simplicity, we will focus on modern digital media formats like DSLR cameras, digital recorders, and software applications.

Video Recording Techniques

Video recording involves capturing moving images via cameras that use sensors or film stock to record visual data. Modern digital video cameras typically offer a wide range of settings that allow creators to manipulate exposure, frame rate, resolution, and depth of field. The main aspects of video recording include:

- **Frame Rate**: Frame rate is the number of individual frames that make up one second of video. Common frame rates are 24 fps (frames per second) for cinematic look, 30 fps for general video, and 60 fps for high-motion content like sports.
- **Resolution**: Video resolution refers to the number of pixels in each frame. The most common resolutions are 1080p (Full HD), 4K (Ultra HD), and increasingly, 8K.
- **Exposure & Lighting**: Proper exposure (how much light hits the camera sensor) and lighting are critical for achieving the desired look. Too much exposure leads to overexposure, while too little leads to underexposure.

- **Focus**: Controlling focus, often through manual adjustment, allows filmmakers to isolate subjects or create blurred backgrounds (bokeh) for creative effects.

Audio Recording Techniques

Audio recording involves capturing sound with microphones and other devices. For high-quality recordings, the following elements are crucial:

- **Microphone Choice**: Different microphones serve different purposes. Condenser microphones are commonly used for studio recordings because of their sensitivity, while dynamic microphones are more rugged and suitable for live performances or noisy environments.
- **Microphone Placement**: The position of the microphone relative to the sound source affects the recording quality. For example, placing a mic too close can result in distortion, while placing it too far can make the audio too quiet.
- **Room Acoustics**: The acoustics of the recording space play a big role in audio quality. Echoes, reverb, and background noise must be controlled to ensure clear, professional audio.
- **Gain Control**: Adjusting the microphone gain (sensitivity) is essential to avoid clipping or distortion, while also ensuring the audio is loud enough for recording.

2. Linear Editing: A Historical Perspective

Linear editing is the traditional form of editing video and audio, where the footage is recorded in a specific sequence and edited in a straight line, typically from start to finish. This was once the standard method used in television and film production, especially before the digital revolution.

Linear Video Editing

In linear video editing, the editor arranges the video clips in a fixed sequence, working with physical tapes or reels. If a mistake was made, the editor had to rewind the tape and cut it again, often in a very manual, painstaking process.

- **Equipment**: Traditional linear editing required specialized equipment such as tape decks, linear editing machines (e.g., the **VTR**), and video mixers.
- **Process**: The editor would assemble footage in a linear order. If a scene needed to be moved, it would require re-recording all subsequent scenes. Editing was usually done in real-time, so making changes was time-consuming and often irreversible.

For example, in a film editing room, editors would physically cut and tape pieces of film together or use "splicing" techniques to join different shots.

Linear Audio Editing

In the context of audio, linear editing is similar in that you work with physical media, often on a multitrack tape recorder. Each change—such as adding, cutting, or altering a sound—required re-recording, and any modifications had to be performed in real time.

- **Process**: Linear audio editors would have to manually cut and paste sections of tape to adjust the timing or content of a track. This made the process of making small tweaks or changes very labor-intensive.
- **Limitations**: Both video and audio were recorded on a continuous, sequential medium, meaning that once a mistake was made, the entire recording sequence would need to be reworked.

Linear editing was efficient in an era before computers but lacked flexibility. As technology advanced, the rise of digital video and audio editing gave way to a new approach—**non-linear editing (NLE)**.

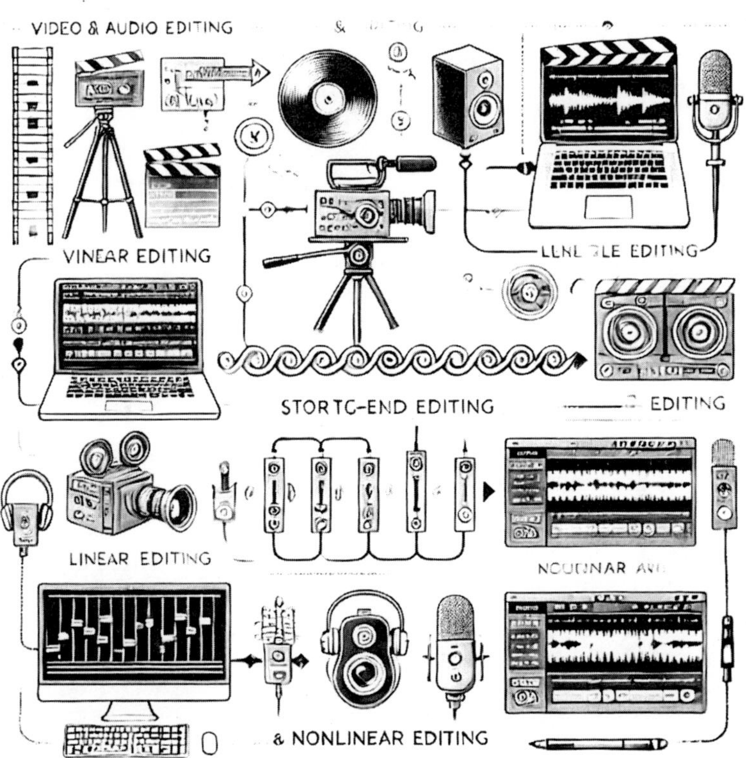

3. Non-Linear Editing (NLE): Revolutionizing the Process

With the advent of computers and digital technology, the film and audio industries transitioned from linear to **non-linear editing** (NLE). NLE allows editors to access any part of the footage or audio at any point in the timeline, without being restricted by the order in which the material was recorded.

Non-Linear Video Editing

Non-linear video editing enables editors to manipulate clips freely without having to work in a sequential manner. This process is often done using software like **Adobe Premiere Pro**, **Final Cut Pro**, or **DaVinci Resolve**, where video clips are stored digitally on hard drives or servers.

- **Digital Files**: Footage is captured in digital formats (e.g., MP4, MOV, or ProRes), making it easy to edit, move, and access clips without worrying about tape or physical media.
- **Timeline-Based Editing**: Editors arrange clips in a virtual timeline, which allows them to quickly move scenes, rearrange shots, and preview changes instantly. The non-linear system also supports multiple tracks for video and audio, allowing for layering, effects, and transitions.
- **Flexibility**: Mistakes can be easily corrected, and edits can be made in any order without needing to re-record or re-shoot.

Non-Linear Audio Editing

Non-linear audio editing operates similarly. It is now primarily done through **DAWs** (Digital Audio Workstations) such as **Pro Tools**, **Logic Pro**, or **Adobe Audition**, where sound files are manipulated on a multi-track interface.

- **Waveform Display**: Audio is displayed as waveforms on the screen, allowing the editor to zoom in and out for precise edits. Editors can cut, move, stretch, or apply effects to any part of the recording at any time.
- **Automation**: NLE systems allow for real-time automation of volume, panning, and effects, enabling the editor to change parameters dynamically as the audio plays.
- **Multi-Track Editing**: Multiple tracks can be combined and edited simultaneously, making complex audio mixing easier and more efficient.

4. Comparing Linear and Non-Linear Editing

Aspect	Linear Editing	Non-Linear Editing (NLE)
Flexibility	Limited, editing must be done sequentially	High flexibility, edit in any order
Editing Process	Time-consuming, requires re-recording	Fast and non-destructive editing
Tools Used	Tape decks, physical film or audio reels	Digital editing software (Premiere, Pro Tools, etc.)
File Management	Linear, continuous footage	Digital clips stored on hard drives or servers
Efficiency	Slower, many manual processes	Faster, with real-time preview and adjustments

Conclusion: Embracing Modern Editing Techniques

The shift from linear to non-linear editing has revolutionized the way we create and edit both video and audio. Non-linear editing allows for far greater creative freedom and efficiency, enabling editors to quickly make changes, try out different ideas, and perfect the final product. Today, video and audio editing are seamlessly integrated into digital workflows, enabling creators across the globe to produce high-quality content more efficiently than ever before.

As technology continues to evolve, new techniques and tools will emerge making the editing process even more accessible and flexible. By mastering both traditional linear techniques and modern non-linear editing methods content creators can ensure they're always ready to adapt to new challenges and create compelling media experiences.

18

Editing of the Recorded Outdoor Programme by Non-linear Editing

Outdoor programming, whether for documentary films, nature shows, travel vlogs, or live-event broadcasts, presents unique challenges and opportunities for video editors. Unlike studio-based productions, outdoor recordings are subject to unpredictable variables like lighting, sound quality, and environmental factors. However, the beauty of non-linear editing (NLE) is that it allows editors to manage these challenges creatively and efficiently. This chapter will delve into the process of editing a recorded outdoor programme using NLE tools, providing an in-depth look at the techniques and considerations that can help shape the final product.

1. The Nature of Outdoor Programmes

Outdoor programmes, by their very nature, capture content in dynamic and uncontrolled environments. This could mean shooting in vast natural landscapes, bustling cities, or remote locations with limited access to professional sound and lighting equipment. As a result, outdoor footage often contains elements that require careful editing to ensure the final product is visually and audibly appealing to the audience.

Some common features of outdoor recordings include

- **Unpredictable Lighting**: Natural light changes constantly due to weather, time of day, and location.
- **Background Noise**: Wind, wildlife, traffic, and crowd noises are common and can make dialogue or key audio hard to hear.
- **Shaky Camera Movements**: Many outdoor programmes involve handheld cameras, drones, or action shots, resulting in unstable footage.
- **Environmental Factors**: Elements like rain, fog, or even dirt on lenses can affect the image quality.

Despite these challenges, NLE tools allow editors to correct or enhance these aspects, transforming raw footage into a compelling narrative.

2. Pre-Editing Preparations: Organizing Raw Footage

Before diving into the editing process itself, one of the first tasks in editing an outdoor programme is to **organize the raw footage** effectively. This will save time during the editing process and ensure that the editor can quickly locate and manage the material they need.

Footage Ingestion

1. Footage Ingestion

Footage ingestion is the process of importing and organizing raw video clips from cameras, storage devices, or online sources into an editing software or digital workflow. It is the first step in any post-production workflow is ingesting the footage into your NLE system. This involves transferring video files from cameras, drones, memory cards, or external hard drives into your computer or NLE software.

Steps in Footage Ingestion

1. **Transferring Files**
    - Connect the storage medium (e.g., SD card, external hard drive) to your computer.
    - Copy footage to a designated folder or project directory.
2. **Organizing Files**
    - Sort files into folders based on:
        - **Date**: E.g., YYYY-MM-DD.
        - **Scene**: Scene 1, Scene 2.
        - **Camera Angles**: Wide shot, close-up, etc.
        - **Takes**: Take 1, Take 2.
        - Use consistent file-naming conventions (e.g., `Scene01_Take02_CamA`).
3. **Backup and Storage**
    - Create at least two backups (e.g., on an external hard drive and cloud storage).
    - Verify all files are transferred correctly before formatting the source device.
4. **Importing into Editing Software**
    - Open your editing tool (e.g., Adobe Premiere Pro, Final Cut Pro, DaVinci Resolve).

- Use the **Media Browser** or drag-and-drop to import footage.
- Maintain the original folder structure for easy navigation.

5. **Metadata Management**:
   - Add tags, keywords, or descriptions to footage for easier searchability.
   - Attach scene numbers, shot types, or other production details.
6. **Proxy Creation** (Optional)
   - Generate low-resolution proxies for smoother editing if working with high-resolution files like 4K or 8K.

File Management: Organize the files by scene, take, and camera angle. Label each clip clearly to avoid confusion later on. NLE programs like **Adobe Premiere Pro**, **DaVinci Resolve**, or **Final Cut Pro** often allow you to create bins or folders to categorize files effectively.

Backup: Always create backups of your raw footage to avoid data loss. You can use RAID drives or cloud storage for redundancy.

Rough Cut

A rough cut is the initial stage of editing, where raw footage is assembled in a sequence without focusing on polishing or fine-tuning. It's about organizing and arranging clips to establish the basic structure of the project.

Steps in Creating a Rough Cut

1. **Review Raw Footage**
   - Watch all footage to familiarize yourself with the material.
   - Mark the best takes or moments using in/out points or color labels.
2. **Create a Timeline**
   - Drag the selected clips into the timeline in the order dictated by your storyboard or script.
   - Focus on arranging scenes logically without worrying about transitions or effects.
3. **Trim and Arrange**
   - Cut excess parts of clips (e.g., mistakes, unnecessary pauses).
   - Arrange clips based on the narrative or sequence flow.
4. **Group Related Clips**
   - Use bins or folders in the software to group clips by scenes, camera angles, or takes.
   - Apply markers to flag important sections within the timeline.

5. **Add Placeholder Elements**

- Insert placeholders for missing footage, graphics, or audio (e.g., "[INSERT VOICEOVER HERE]").
- Add basic title cards or labels for scenes if needed.

6. **Sync Audio**
   - If external audio was recorded, sync it with the video using clapperboard marks or waveform matching.
   - Ensure audio levels are clear for reviewing purposes.
7. **Basic Sequence Setup**:
   - Arrange clips into acts, scenes, or chapters.
   - Ensure no major gaps or overlaps disrupt the flow.
8. **Review and Iterate**
   - Share the rough cut with collaborators for feedback.
   - Make necessary changes to the structure or scene order.

Once the footage is organized, it's time to start creating a rough cut. A rough cut involves laying down the footage in a timeline without worrying too much about fine details. This is where the overall structure of the programme begins to take shape.

- **Timeline Setup**: Set the timeline according to the project's specifications (e.g., 1080p, 4K, 24fps, etc.). It's important to match the timeline to the footage format to avoid confusion later.
- **First Pass Editing**: Begin by placing clips in chronological order, following the general flow of the script or planned sequence. The goal at this stage is not to make the footage perfect but to get a basic sense of the programme's narrative.

Syncing Audio and Video

Often, outdoor productions involve separate audio and video recordings (for instance, using a shotgun mic or lapel mic while recording the video with a camera). NLE tools are designed to sync audio and video tracks effortlessly.

- **Auto-Syncing**: Most modern NLE software has auto-sync features that can align audio with video based on timecode, waveform analysis, or clapperboard markers.
- **Manual Syncing**: If auto-sync doesn't work perfectly, editors can manually sync audio by matching visual cues (like claps or visible movements) to the corresponding sound in the audio track.

3. Fine-Tuning the Visuals: Color Correction and Stabilization

Once the rough cut is assembled, editors begin to polish the footage to create a professional, cinematic look. The key areas to focus on for outdoor programmes are **color correction**, **color grading**, and **stabilization**.

Color Correction

Outdoor footage often suffers from inconsistent lighting due to varying conditions like overcast skies, sunset lighting, or shadowed areas. Color correction addresses these issues and ensures the footage is visually consistent.

- **White Balance**: Outdoor lighting can often cause a color cast in the footage (e.g., too warm during golden hour or too cool under cloudy skies). Adjusting the white balance in the NLE helps ensure natural-looking colors.
- **Exposure**: Outdoor shoots might include both very bright and very dark areas. Use the NLE's color grading tools to adjust the exposure, lifting shadows and controlling highlights to ensure details are visible.
- **Correcting Skin Tones**: One of the most important tasks is making sure that skin tones look natural across different lighting environments. In NLEs like DaVinci Resolve, you can use the color wheels or the curves tool to make subtle adjustments.

Color Grading

After basic color correction, color grading enhances the aesthetic mood of the programme. Editors might want to give the footage a "warmer" look for a sunset scene or a more "desaturated" appearance for a gritty documentary.

- **Creative LUTs (Look-Up Tables)**: Many NLEs provide built-in LUTs that apply predefined color looks (e.g., cinematic, vintage, etc.) to the footage. These LUTs can serve as starting points for grading.
- **HSL Adjustment**: Fine-tune individual colors (e.g., enhancing the green of the trees or the blue of the sky) to create visual harmony.

Stabilization

Outdoor footage often includes handheld shots or scenes with a moving camera (e.g., walking or running), leading to shaky visuals. Stabilizing these shots can improve the viewer's experience.

- **Stabilization Tools**: Modern NLE software comes with stabilization features that automatically reduce shake. In DaVinci Resolve, for instance, you can apply the "Zoom" or "Cropping Ratio" stabilization modes to smooth out shaky clips.

- **Manual Stabilization**: For particularly difficult shots, editors can manually adjust the position of the footage in the frame, compensating for movement or shake.

4. Enhancing the Audio: Noise Reduction and Sound Design

In outdoor environments, audio recording can be challenging due to wind, crowd noise, traffic, and other environmental factors. Fortunately, NLE tools offer advanced audio editing and enhancement options to ensure clarity.

Noise Reduction

Outdoor recordings often contain unwanted background noise, such as wind, rustling leaves, or distant conversations. NLE software can help isolate and reduce this noise.

- **Noise Profiles**: Use tools like **Adobe Audition**'s Noise Reduction or **iZotope RX** to create noise profiles. These tools analyze the background noise and then remove it from the entire track.
- **High-Pass Filters**: Applying a high-pass filter removes low-frequency sounds like hums or wind noise, leaving the clearer frequencies intact.

Audio Repair

In some cases, you might need to repair damaged or distorted audio, such as clipping caused by sudden loud noises.

- **De-clip**: Many NLEs offer a de-clipping feature to recover some of the audio data from overdriven sounds.
- **EQ and Compression**: Equalization (EQ) can be used to adjust the tonal balance of the audio, while compression ensures the sound levels are consistent throughout the programme.

Sound Design

Sound design can elevate outdoor content by enhancing the natural atmosphere of the environment or adding sound effects to intensify the narrative.

- **Foley Effects**: Record custom sound effects or add stock sound effects that complement the visuals (e.g., footsteps in the forest, rustling leaves, or water flowing).
- **Ambient Sound**: Layering ambient sound (e.g., birds chirping, wind rustling) beneath the dialogue can create a more immersive experience for the viewer.

5. Finalizing the Programme: Transitions, Titles, and Exporting

With the video and audio polished, it's time to add the finishing touches. This includes adding transitions, titles, music, and any final adjustments before exporting.

Transitions

Transitions like fades, dissolves, or wipes can be used to seamlessly move between scenes or locations. For outdoor programmes, slow crossfades between scenes can create a smooth flow that mirrors the natural progression of time.

Titles and Graphics

Adding titles (such as episode names or location identifiers) and graphics can give context to the viewer. For example, showing the name of the location in a travel programme or the species of animals in a wildlife show.

Exporting

Once the final edit is complete, it's time to export the programme for distribution. Choose the appropriate export settings based on the intended platform (e.g., 1080p for YouTube, 4K for film festivals). Most NLEs offer export presets for different platforms to simplify this process.

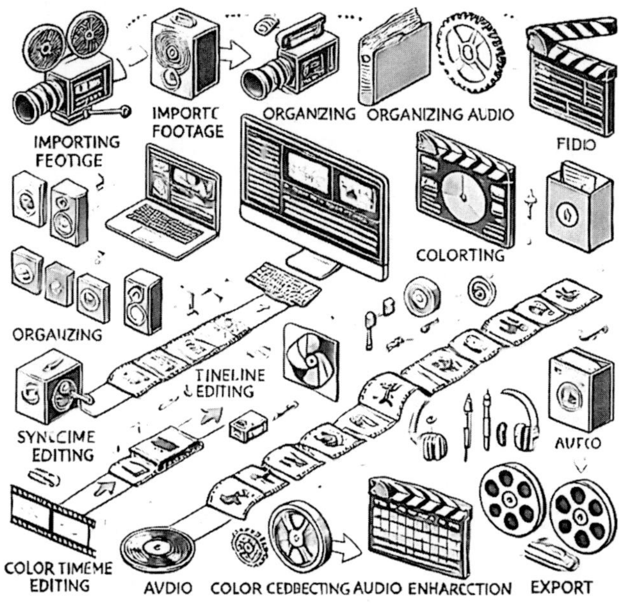

Image: Workflow of non-linear editing for an outdoor recorded program

Advanced Techniques for Enhancing Outdoor Footage

1. Multi-cam Editing

When shooting from different angles or cameras, multi-cam editing allows you to seamlessly switch between perspectives, adding depth and variety to the scene. This feature is particularly useful for event coverage, interviews, or wildlife footage, where changing viewpoints enhances the viewer experience.

2. Motion Tracking and Stabilization

Unsteady footage, common in outdoor programs, can be stabilized in post-production. Some NLE platforms offer motion tracking to smooth out movement, ensuring a steady frame that doesn't distract the viewer. This is especially useful for action shots or scenes recorded with handheld cameras.

3. Slow Motion and Time-lapse

Outdoor programs can benefit from time manipulation effects. Slow-motion captures intricate details, while time-lapse condenses prolonged activities into short, compelling visuals. These effects are excellent for highlighting nature, events, or other dynamic outdoor elements.

4. Green Screen and Compositing

Though not always required, compositing can add additional layers to the video. For instance, if a presenter is shot against a green screen, background footage from the actual outdoor location can be added later. Compositing provides flexibility in cases where retakes in the outdoor setting aren't feasible.

Optimizing the Editing Workflow for Outdoor Programs

NLE software provides various features that help maintain an organized and efficient workflow:

- Proxy Editing for Large Files: Outdoor shoots typically involve high-resolution footage, which can slow down the editing process. Working with proxy files — lower-resolution versions of the raw footage — can make the workflow smoother, especially on lower-spec hardware.
- Keyboard Shortcuts and Customization: Most NLE platforms offer customizable keyboard shortcuts. Configuring these shortcuts for frequently-used commands, like cutting, splicing, and playback controls, can significantly speed up the editing process.
- Asset Management and Archiving: Outdoor footage often takes up significant storage space. Properly label and store files for easy access, and once the project is completed, archive the assets to free up storage.

Conclusion: The Power of Non-Linear Editing for Outdoor Programmes

Non-linear editing has revolutionized the way editors work with outdoor recordings. The ability to manipulate video and audio in flexible, non-destructive ways provides editors with powerful tools to overcome the challenges of outdoor production. By mastering NLE techniques such as color correction, stabilization, sound design, and noise reduction, editors can transform raw, chaotic footage into a cohesive, engaging programme.

In the world of outdoor programming, the unpredictability of nature can never be fully controlled, but non-linear editing allows editors to work with this unpredictability and create polished, professional content that captivates audiences worldwide.

19

Importing Video: Working Methods of Offline Video Editing

In modern video production, one of the most critical steps in the post-production workflow is the importation of video. This process marks the transition from raw footage to a structured project that can be edited, refined, and turned into a finished product. Effective offline video editing techniques, which allow for the editing of low-resolution proxies of the original footage, are crucial for enhancing workflow efficiency, particularly in large-scale or resource-heavy projects.

This chapter will guide you through the fundamental working methods of offline video editing, focusing on the process of importing video footage into your editing system, preparing that footage for editing, and understanding how offline editing fits into the overall post-production pipeline.

1. Understanding the Workflow of Offline Editing

Offline video editing is a method used to edit a project using lower-resolution versions (or proxies) of the original high-resolution video footage. This method allows editors to work faster, save on computer processing power and storage space, and still be able to create a solid edit that can later be "conformed" to the original high-resolution footage in the final stage of post-production (referred to as "online editing").

Offline vs. Online Editing

- **Offline Editing**: The editing is done on proxy files (lower-resolution copies of the original footage) which are easier to handle by the system. Offline editing is primarily focused on crafting the narrative structure of the project—making cuts, arranging clips, and working on timing and flow.
- **Online Editing**: Once the edit is locked in the offline stage, online editing begins. This process involves replacing the proxy footage with the original high-resolution files, along with final color grading, audio mixing, and any visual effects or enhancements that require the highest quality possible.

The offline editing workflow, therefore, allows for more flexibility and speed while not sacrificing the final quality of the project, as it facilitates the seamless transition to high-quality media during the online editing process.

2. Importing Video Footage into the Editing System

The first step in any video editing workflow is importing the footage into the editing system. The method of importation can vary depending on the type of footage, the camera used to record it, and the software being used for editing.

Types of Video Formats

Before importing video into an NLE (non-linear editing) system, it's essential to understand the different types of video formats, as each format requires specific handling.

- **Uncompressed Formats** (e.g., **ProRes 422**, **DNxHD**, **AVC-Intra**): These formats retain the highest quality of the video but require large file sizes, making them more difficult to work with in offline editing.
- **Lossy Formats** (e.g., **H.264**, **H.265**, **MP4**): These formats are compressed and reduce file sizes significantly, making them more manageable for offline editing.
- **Raw Footage** (e.g., **ARRIRAW**, **RED R3D**, **Blackmagic RAW**): Raw footage from high-end cameras retains a vast amount of information and dynamic range, allowing for more flexibility in color grading but can be quite heavy on storage.

Import Process in NLEs

Each NLE software (e.g., **Adobe Premiere Pro**, **Final Cut Pro X**, **Avid Media Composer**, **DaVinci Resolve**) has its method of importing footage, but the general process involves:

- **Connecting Media Sources**: This includes ingesting footage from memory cards, hard drives, cameras, or cloud-based storage.
- **File Management**: Once the media is connected, video files should be organized logically. Many editors prefer creating folders or bins based on the type of shot, scene, or camera angle.
- **Choosing Import Settings**: The NLE software allows editors to configure how files are imported. This includes setting the import location, creating proxies, or choosing to copy or link to the media.
- **Linking to Original Files**: In most modern editing workflows, NLEs don't move or alter the original video files; instead, they link to them. This process ensures that the editor is always working with the same files but only sees references to them within the software.

For example, in **Adobe Premiere Pro**, media can be imported by using the **Media Browser** or **Import** dialog box. Premiere allows users to link directly to camera files or transcode them into a preferred editing format.

3. Creating Proxies for Offline Editing

In an offline editing workflow, particularly for large video files such as 4K or 8K footage, editors often work with **proxy files**—low-resolution versions of the original media files. This practice allows for smoother editing performance and a much more efficient workflow without needing high-powered systems.

Why Use Proxies?

- **Faster Playback**: High-resolution files can slow down the editing process due to their large size. Proxies enable real-time playback and faster scrubbing through timelines.
- **Reduced Storage Requirements**: Working with proxies minimizes the need for large storage spaces, allowing editors to store and work with multiple projects more easily.
- **Better System Performance**: Especially when working with older or less powerful hardware, proxies allow editors to work without being bogged down by the demands of processing full-resolution video.

Creating Proxies in NLEs

Many modern NLE systems have built-in tools for generating proxies during the import process:

- **Adobe Premiere Pro**: Premiere Pro offers an easy-to-use proxy workflow that allows users to ingest footage and automatically generate proxies. When importing media, editors can choose to create proxies by selecting an ingest preset that transcodes the media to a smaller, easier-to-edit format like **ProRes Proxy** or **H.264**. Later, the system will automatically relink to the high-res media during the online process.
- **Final Cut Pro X**: FCPX offers proxy workflows by transcoding high-resolution media into **ProRes Proxy** or **H.264** at lower resolutions like 720p. This proxy workflow is fully integrated with the app's Media Browser, making it easy to switch between proxies and full resolution for editing.
- **Avid Media Composer**: Avid allows users to create proxies during the import phase, or **Transcoding**, by selecting a lower-resolution codec for editing. Avid's powerful media management system ensures the proxies are seamlessly relinked to the high-resolution files during the conform stage.

Proxy Workflows

The general proxy workflow looks like this:

1. **Ingest**: Import footage into the NLE and create proxy files.
2. **Edit**: Work with the proxy files in the timeline. Since these files are lower resolution, the editing process will be faster.
3. **Relink**: Once the rough cut is finalized, the proxies are relinked to the original high-res footage.
4. **Online Editing**: The project is now ready for online editing, where color correction, audio mixing, and visual effects are applied to the high-res files.

4. Key Considerations for Offline Editing

While offline editing with proxies is common, there are several factors to keep in mind when working with this approach:

1. Quality Control

When working with proxies, it's important to ensure that the proxies are close enough in appearance to the final high-resolution footage to avoid any surprises during the final conform. Always check that the proxies are accurate in terms of framing, exposure, and color before committing to final edits.

2. Media Management

Proper media management is crucial for offline editing. You should ensure that all media is carefully organized, named, and stored in a way that makes it easy to locate throughout the editing process. Using bins or folders that are clearly labeled by scene, shot type, or take is highly recommended.

3. Storage and Backup

During the offline editing process, you'll be working with large amounts of data. Always ensure that the proxies, as well as the original footage, are backed up regularly. Using a RAID storage system or cloud-based backups can help safeguard against data loss.

4. Audio Synchronization

When using proxies, you'll often need to sync audio recorded separately from the video. Many NLEs, including **Adobe Premiere Pro**, **Final Cut Pro**, and **DaVinci Resolve**, have tools to automatically sync audio to video by matching audio waveforms or timecode.

Conclusion: The Power of Offline Video Editing

Importing video and using offline editing workflows is a highly efficient way to work with large-scale video projects. By creating proxies, editors can streamline their workflow, make real-time decisions about pacing and structure, and collaborate more effectively across teams.

By understanding how to manage media, create proxies, and implement offline video editing methods in NLE systems, editors can maximize their efficiency and focus on crafting compelling stories with their raw footage. The offline editing phase is just the beginning, but with a proper foundation in place, it sets the stage for a smooth and successful transition to online editing and the final production.

20

Advanced Techniques of Editing Cuts, Mixes, and Advanced Software

As video editing continues to evolve, so do the techniques and software that enable editors to create polished, captivating visuals. Advanced editing techniques like precision cutting, mixing, and using sophisticated software tools allow editors to enhance storytelling, manipulate pace, and shape the emotional experience of their audience. This chapter dives into these advanced techniques, exploring how each tool and method contributes to a compelling final product.

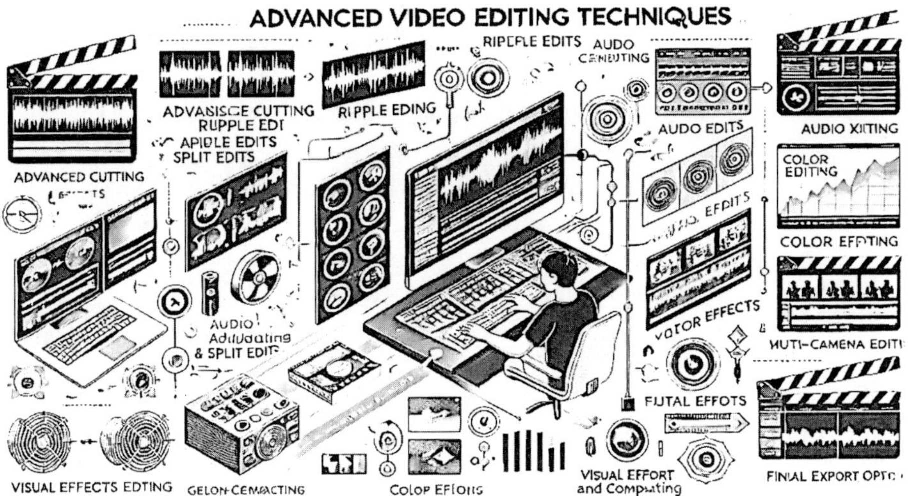

1. The Art of Cutting: Techniques for Seamless Storytelling

Cutting is a fundamental editing skill that determines the pacing, continuity, and flow of a video. Advanced cutting techniques, such as L-cuts, J-cuts, and match cuts, serve as the foundation for creating dynamic narratives.

- Types of Cuts
    - L-Cut and J-Cut: L-cuts and J-cuts are used to transition between audio and video in ways that flow more naturally and create smoother

continuity between scenes. These are essential techniques for dialogue-driven scenes or scenes that require seamless transitions between different locations or times.

- **L-Cut**: In an L-cut, the audio from the next scene starts before the video does, allowing the sound to create a bridge between the two scenes.
- **J-Cut**: In a J-cut, the video starts before the audio does, often used to transition between locations or times with a slight delay in the audio.

Both cuts are typically used to create a smoother narrative experience, allowing the editor to control the rhythm of the conversation and match visual elements more effectively with audio cues.

- Match Cut: A match cut connects two scenes that share visual or thematic elements, creating a smooth transition between them. This can be done by matching shapes, movements, or even sounds between the two shots.
  - **Purpose**: The match cut works as a bridge, creating continuity between two disparate scenes. It can be used to emphasize visual or thematic motifs, such as a transition from one setting to another or to symbolize a character's evolution.
  - **Example**: Stanley Kubrick's famous **match cut** from a bone thrown into the air (in *2001: A Space Odyssey*) to a spacecraft in orbit is an iconic example, illustrating humanity's leap from prehistory to space exploration.
- Jump Cut: The jump cut is one of the most creative and dynamic editing techniques, often used in modern filmmaking and YouTube-style videos. This cut involves the removal of a section of time from a continuous shot, creating a noticeable "jump" in the action.
  - **Purpose**: A jump cut can show a significant passage of time, accelerate the pacing, or highlight the importance of a specific moment. It can also add a feeling of disorientation or surprise when used unexpectedly.
  - **When to Use**: Jump cuts are often used in vlogs, montages, or situations where you want to keep the audience engaged without long, unbroken takes. However, jump cuts can also disrupt the flow of traditional narrative filmmaking if not executed carefully.

Cross-Cutting (Parallel Editing)

Cross-cutting, also known as parallel editing, involves cutting between two or more scenes happening simultaneously but in different locations. This technique is often used to build tension or show the interconnection between characters and events.

- **Purpose**: It is commonly used in action sequences or thrillers to heighten suspense or to show multiple events occurring at once that will eventually converge in the narrative.
- **Example**: In **The Godfather** (1972), cross-cutting is used during the baptism scene, where Michael Corleone is at the church, while his men are carrying out the murders. The simultaneous actions intensify the narrative and thematic tension.
- Cut Timing and Precision

Achieving the right timing in cuts can significantly impact the storytelling:

- Pacing Adjustments: Slow-paced cuts give viewers time to absorb detail, while quick cuts can heighten tension or energy.
- Cutting on Action: Cutting at the peak of an action, such as a door closing or a person turning, creates smoother transitions and keeps the visual flow uninterrupted.
- Emotional Beats: Timing cuts with emotional peaks in music or dialogue enhances the audience's connection to the story.

2. Mixing Techniques for Visual and Audio Cohesion

Mixing refers to the blending of audio and visual elements to create a unified experience. Advanced mixing techniques, including audio equalization, multi-layered sound design, and the use of color grading software, contribute to an immersive viewer experience.

- Audio Mixing Techniques

Audio plays an equally important role in video editing as the visual elements. Advanced audio mixing techniques ensure that the soundtrack, dialogue, sound effects, and ambient sounds work together seamlessly to enhance the narrative. Audio is as crucial as visuals in video editing, and advanced audio mixing ensures clarity, mood, and emphasis.

Audio Sweetening

Audio sweetening refers to the process of improving the quality of the recorded audio. This involves cleaning up background noise, enhancing clarity, and adding depth to the sound.

- **Noise Reduction**: Using tools like **iZotope RX** or **Adobe Audition** to remove unwanted background sounds (e.g., hums, buzzes, traffic noise, etc.) that detract from the clarity of the dialogue or music.
- **EQ**: Equalization (EQ) is essential for balancing frequencies, making the dialogue clear, and ensuring the music and sound effects don't

overpower the spoken word. High-pass filters can eliminate low-end rumble, while boosting mid and high frequencies can make vocals and instruments pop.

ADR (Automated Dialogue Replacement)

ADR is a technique used to re-record dialogue in a controlled environment after the original sound recording, typically due to poor audio quality or additional dialogue required for clarity.

- **Purpose**: ADR is crucial when the original on-location audio is unusable due to noise, microphone issues, or technical failures.
- **Process**: In a typical ADR session, actors watch the video playback and synchronize their re-recorded lines to match the lip movements on-screen.

Surround Sound Mixing

For films or projects with a complex audio environment, editors can mix audio in **surround sound** (e.g., 5.1 or 7.1). This involves positioning sound elements in different channels (front, rear, center, subwoofer) to create a more immersive experience.

- **Software**: Editors use software like **Avid Pro Tools** or **Adobe Audition** to create surround sound mixes. This technique is essential for creating the immersive environment that pulls the audience into the narrative.
    - Equalization (EQ): Adjusting frequency levels to emphasize specific audio elements. For example, lowering low frequencies can create clarity in dialogue, while boosting mids can enhance ambient sounds.
    - Layering Sound Effects: Background sounds (e.g., city traffic, forest sounds) can layer with dialogue to add context. Multi-track editing allows for separate manipulation of dialogue, music, and sound effects.
    - Audio Ducking: This technique lowers the volume of background sounds when primary audio, such as dialogue, is playing, ensuring the key sounds are prominent.
- Visual Mixing Techniques

Advanced visual mixing involves color grading, overlays, and blending modes to achieve a cohesive look.

- Color Grading: Software like DaVinci Resolve allows for precise adjustments to color, saturation, and contrast, setting the mood for the entire project.

- Overlay Blending: Blending modes like multiply, overlay, or screen can mix visuals, creating effects like light leaks or textured backdrops. This technique is widely used in music videos and dramatic scenes.
- Split-Screen and Multicam: Split-screens allow for simultaneous storytelling by placing multiple shots side-by-side, ideal for dialogue scenes, phone calls, or showcasing contrasting actions.

3. Advanced Software for Precision and Creativity

With modern editing software, editors have access to advanced tools for every stage of the editing process, from cuts to final mixing.

- Non-Linear Editing Software (NLE)

NLEs like Adobe Premiere Pro, Final Cut Pro, and DaVinci Resolve are popular due to their flexibility and control over cuts and mixes. Key features include:

- Multi-Cam Editing: This feature allows editors to sync and cut between multiple camera angles seamlessly, providing a dynamic and versatile editing experience.
- Dynamic Link: Adobe's Dynamic Link, for instance, allows smooth transitions between Premiere Pro and After Effects without exporting files, preserving quality and allowing real-time editing.
- Keyframe Animation: This technique animates effects or transitions over time by setting points (keyframes) where specific parameters, such as opacity or scale, change.

- Audio Editing and Mastering Software

Specialized audio editing tools can significantly enhance audio quality:

- Adobe Audition: Provides noise reduction, reverb adjustment, and multi-track mixing. Audition's spectral frequency display makes it easy to isolate and remove unwanted sounds.
- Pro Tools: Widely used in professional studios, Pro Tools is favored for its detailed control over audio tracks, effects, and mastering, ideal for projects with complex sound requirements.
- Logic Pro: Known for its intuitive interface and high-quality plug-ins, Logic Pro allows editors to create custom soundtracks and apply effects like reverb, echo, and distortion.

4. Innovative Editing Techniques with Advanced Software

Advanced software allows editors to experiment with unconventional editing techniques, adding layers of creativity to the final production.

- Motion Graphics and Visual Effects

Software like Adobe After Effects enables the creation of stunning visual effects and motion graphics:

  - Compositing: Combining elements from multiple sources to create a single, unified scene. Compositing is frequently used for green-screen effects.
  - 3D Camera Tracking: Allows for 3D effects that move with the camera, adding realistic depth to scenes.
  - Particle Systems: Used to create effects like snow, rain, or explosions, enhancing action scenes or adding a dramatic touch.

- Non-Destructive Editing

Non-destructive editing allows for flexible experimentation without affecting the original footage:

  - Adjustment Layers: Non-destructive adjustments in color, exposure, or effects can be applied across clips without altering the base footage.
  - Proxy Editing: Editing low-resolution proxy files speeds up workflow and allows for seamless transitions to high-resolution files for final export.

- AI-Assisted Tools and Automation

AI-powered tools are revolutionizing video editing by automating complex processes:

- Auto Reframe: AI detects important areas of a shot and reframes automatically, saving time when converting videos for multiple formats.
- Speech-to-Text: Automated transcription tools speed up captioning and subtitles, improving accessibility.
- Face Tracking and Scene Recognition: AI algorithms track faces or specific objects within footage, enabling precise adjustments to those elements throughout the edit.

Conclusion

Advanced editing techniques—whether it's applying creative cuts, refining audio, or leveraging the capabilities of modern editing software—play a key role in producing high-quality video content. By mastering techniques like jump cuts, L-cuts, cross-cutting, and complex audio mixing, editors can elevate their storytelling, guiding the viewer's emotions and attention with precision.

Moreover, the integration of powerful editing software like DaVinci Resolve Adobe Premiere Pro, Avid Media Composer, and Final Cut Pro X unlocks possibilities that were once unimaginable. These tools allow editors to experiment with complex visual effects, advanced color grading, precise audio mixing, and efficient workflows, empowering them to bring their creative visions to life.

By combining technical skill with creativity, editors can produce polished professional work that is visually engaging, emotionally resonant, and narratively compelling.

21

Using Graphics and Animation in Video Production

In today's fast-paced visual media landscape, the integration of graphics and animation has become a crucial element in storytelling, branding, and information delivery. Whether used for subtle enhancements or to drive the visual narrative, these tools help to elevate a video from simple footage to a compelling visual experience. This chapter explores the role of graphics and animation in video production, offering a guide to their use, techniques, and best practices in creating effective and engaging content.

1. The Role of Graphics and Animation in Video Production

Graphics and animation serve multiple functions in modern video production, adding both style and substance to the final product. They can be utilized to clarify complex information, deliver branding messages, provide visual transitions, or create entirely immersive experiences.

1.1 Informational Graphics and Data Visualization

One of the most prevalent uses of graphics is in the representation of data. **Infographics** and **data visualizations** transform raw statistics into engaging visuals that are easier to digest and more appealing to viewers.

- **Purpose**: To present complex data and concepts in a simple, visually engaging way. Think of corporate videos, educational videos, or documentaries where a viewer may need to understand statistics, trends, or comparisons.
- **Tools Used**: Programs like **Adobe After Effects**, **Apple Motion**, and **Veed.io** offer tools to create interactive graphics, charts, and animated diagrams. The use of motion in these graphics (e.g., bar graphs animating in) can enhance viewer engagement and retention.

1.2 Branding and Logo Animation

Incorporating animated graphics is an effective way to establish or reinforce a brand identity. Animated logos, titles, and transitions play a vital role in giving a video its professional look while also ensuring that a brand message is communicated.

- **Purpose**: To create a strong, memorable brand presence. For instance, animated logo reveals at the beginning or end of a video help reinforce brand recall.
- **Examples**: The opening sequence of films or TV shows often includes a creative logo animation. Similarly, YouTube channels frequently use custom logo animations to brand their content.

1.3 Storytelling Through Animation

Animation is not just for explainer videos or branding—animation can be a powerful tool in creative storytelling. Entire films and series, such as those created by **Pixar** or **Studio Ghibli**, rely heavily on animation to build immersive worlds and tell compelling stories.

- **Purpose**: Animation allows for boundless creativity, enabling filmmakers to build imaginative worlds, create fantastical creatures, or visualize abstract concepts. In video production, animation can serve as a medium for storytelling in itself or as a supplement to live-action footage.
- **Examples**: Hybrid formats where live-action footage is enhanced with animated elements, such as **"Who Framed Roger Rabbit"** (1988) or **"Space Jam"** (1996), illustrate how seamlessly animation can be woven into the narrative of a film or commercial.

1.4 Motion Graphics for Transitions and Effects

Motion graphics are typically used for transitions, lower-thirds (on-screen text that identifies people or places), titles, and other visual effects that help polish the overall look of the video.

- **Purpose**: To ensure smooth transitions between scenes or to highlight key information in a visually dynamic way. Motion graphics make the viewing experience more engaging by keeping the visuals active and the pacing tight.
- **Examples**: Think of **lower-third graphics** introducing speakers in interviews, **title sequences** that introduce chapters or segments, or **animated transitions** that move between scenes with fluidity.

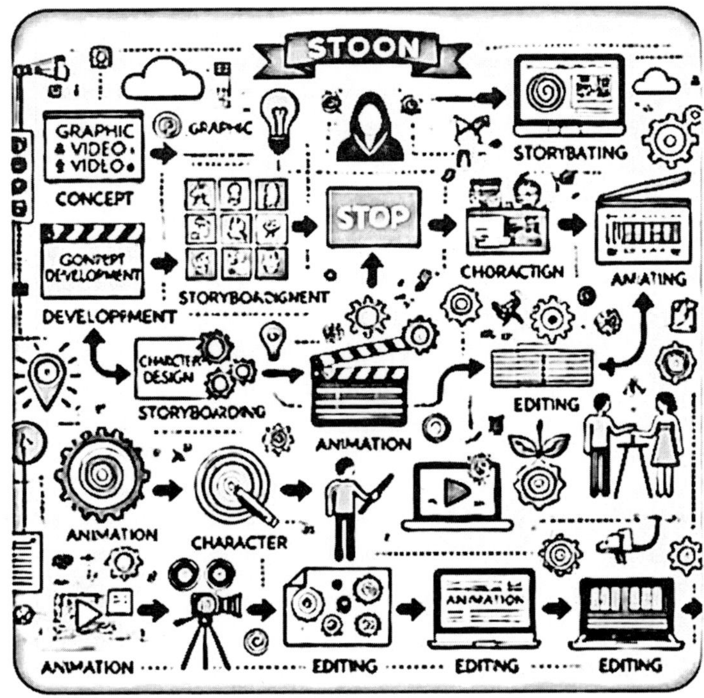

2. Key Techniques in Graphics and Animation for Video

The creation of effective graphics and animation involves a combination of artistic skill, technical expertise, and creative thinking. There are several advanced techniques used in professional video production to incorporate graphics and animation.

2.1 Keyframing and Animation Curves

Keyframing is the process of setting specific points in time (keyframes) that define changes in motion or visual properties (position, scale, opacity, etc.). This is the foundation of animation in video production, allowing for the creation of fluid motion by adjusting the movement of objects between these points.

- **Purpose**: To bring static graphics to life by creating movement. Keyframing is the most common method of animating graphics, shapes, and text in programs like **After Effects** or **Cinema 4D**.
- **Example**: An animated logo might start at a small size and gradually zoom in to full size, while text may move across the screen, fading in and out. The smoothness of these animations is controlled through keyframes and animation curves (e.g., ease-in and ease-out functions).

2.2 3D Animation and Modeling

While 2D animation is widely used in video production, the integration of **3D animation** and **modeling** has become increasingly common. This includes creating 3D objects, characters, and environments that are then animated within a three-dimensional space.

- **Purpose**: To produce more complex, realistic, or fantastical animations that add depth and a sense of immersion. 3D animation is used in a wide range of contexts, from visual effects in movies to product demonstrations and virtual reality (VR).
- **Tools Used**: **Autodesk Maya**, **Blender**, and **Cinema 4D** are powerful tools for creating 3D models and animations. In combination with software like **Adobe After Effects**, 3D elements can be integrated into live-action video footage for added realism.

2.3 Typography and Kinetic Typography

Kinetic typography is a technique where text is animated in a way that conveys meaning or emotion. It involves moving, resizing, or changing the style of text in a dynamic manner that complements the visual narrative.

- **Purpose**: Kinetic typography adds energy and emphasis to specific words or phrases, often used in music videos, commercials, or promotional material to convey a message with urgency or impact.
- **Example**: Music videos with lyrics moving to the beat of the music or commercials that highlight key phrases with animated text effects.

2.4 Visual Effects (VFX) and Integration

Visual Effects (VFX) refer to the integration of computer-generated imagery (CGI) into live-action footage. This can include creating explosions, weather effects, digital environments, or even entire characters that do not exist in the physical world.

- **Purpose**: VFX can create visually spectacular scenes that would be impossible, impractical, or expensive to achieve with traditional methods. In addition, VFX can be used to enhance or change the existing footage.
- **Example**: In blockbuster films like **"Avengers: Endgame"**, VFX is used extensively to create entire worlds, characters (e.g., CGI Thanos), and dramatic action sequences.

3. Software for Creating Graphics and Animation

The tools you use for graphics and animation play a crucial role in determining the quality and efficiency of your workflow. Here's a look at some of the most popular software used in the video production industry.

3.1 Adobe After Effects

Adobe After Effects is one of the most widely used tools for creating motion graphics and visual effects. Its versatility and extensive plugin library make it the industry standard for motion designers.

- **Capabilities**: After Effects offers tools for 2D animation, 3D compositing, text animation, and advanced VFX. The software allows animators to work with keyframes, masks, particle systems, and other advanced features.
- **Example**: After Effects is used to create everything from animated titles in films to complex animated infographics.

3.2 Cinema 4D

For 3D graphics and animation, **Cinema 4D** is one of the most popular tools, particularly when paired with After Effects for seamless integration.

- **Capabilities**: Cinema 4D excels at 3D modeling, texturing, rendering, and animating. Its user-friendly interface and robust animation tools make it ideal for motion graphics designers who want to incorporate 3D elements into their work.
- **Example**: Motion designers often use Cinema 4D to create animated 3D logos, product animations, and visual effects.

3.3 Blender

Blender is a powerful open-source 3D creation suite that includes everything from modeling and texturing to animating and rendering.

- **Capabilities**: Blender is known for its ability to handle complex 3D tasks, including sculpting, simulation, compositing, and video editing. Despite being free, Blender is widely used in professional environments due to its depth and flexibility.
- **Example**: Blender can be used to create animated short films, 3D product demonstrations, or even character animations.

3.4 Adobe Illustrator and Photoshop

Although primarily used for still imagery, **Adobe Illustrator** and **Photoshop** are also essential in the creation of assets for animation and motion graphics.

- **Capabilities**: Illustrator is often used to design vector-based elements, while Photoshop is used to create textures or photo-realistic elements. These assets can then be imported into After Effects or other software for animation.
- **Example**: Designers use Illustrator to create vector-based icons and logos, which are then animated in After Effects for use in videos.

4. Best Practices for Using Graphics and Animation in Video Production

4.1 Keep It Relevant

Graphics and animations should always serve the story, not distract from it. Whether you're creating an animated logo or an infographic, ensure that your design enhances the message or tone of the video.

4.2 Focus on Timing and Pacing

The timing of your graphics and animation is just as important as their design. Too fast, and they might overwhelm the viewer; too slow, and they can lose their impact. Experiment with the timing to ensure the animations align with the video's pacing and rhythm.

4.3 Consistency in Style

Whether using 2D or 3D elements, maintaining a consistent style across all graphic elements helps to keep the production cohesive. The choice of colors, typography, and animation style should align with the branding or tone of the video.

4.4 Test for Accessibility

Ensure that any text or graphic elements are legible and accessible to all viewers, including those with visual impairments. Use readable fonts, clear contrast between text and background, and avoid overcrowding the screen with too many moving elements.

Conclusion

Graphics and animation are not merely decorative in video production—they play a fundamental role in conveying ideas, engaging viewers, and enhancing the narrative. Whether for informational purposes, branding, or creative expression, understanding how to effectively integrate these elements into your videos is crucial for any modern content creator. By mastering the techniques of animation, utilizing the right tools, and following best practices, you can take your video production to new heights of creativity and professionalism.

22

Application of Open Source Software Gimp, Blender

Introduction

Open-source software has revolutionized the way individuals and industries approach digital content creation, offering high-quality, community-driven solutions without the cost of proprietary software. Among the most popular open-source tools for graphics and 3D modeling are GIMP (GNU Image Manipulation Program) and Blender. Both programs provide powerful capabilities for creating, editing, and enhancing visual content. This chapter delves into the application of GIMP and Blender, exploring their key features, workflows, and how they can be utilized across various fields of digital production. This chapter explores the capabilities and applications of two prominent open-source software tools: GIMP (GNU Image Manipulation Program) and Blender. These software tools are transformative in the fields of digital design, animation, and media production. The chapter provides an overview of their features, real-world applications, and advantages over proprietary alternatives. Through practical case studies, the chapter demonstrates how GIMP and Blender empower professionals, educators, and hobbyists to create high-quality visual content without incurring significant financial costs.

What is Open-Source Software?

Open-source software is freely available for use, modification, and distribution. It fosters collaboration among developers and users, resulting in cost-effective, versatile, and frequently updated tools.

Why Open Source?

- Cost-effectiveness
- Customizability
- Strong community support
- Access to frequent updates

Overview of GIMP

GIMP is a versatile open-source software that has been widely adopted for its robust image-editing capabilities. Comparable to Adobe Photoshop, GIMP allows users to manipulate images, create artwork, and perform complex graphic design tasks. Developed and maintained by a community of contributors, it supports various plugins and scripts to enhance its functionality.

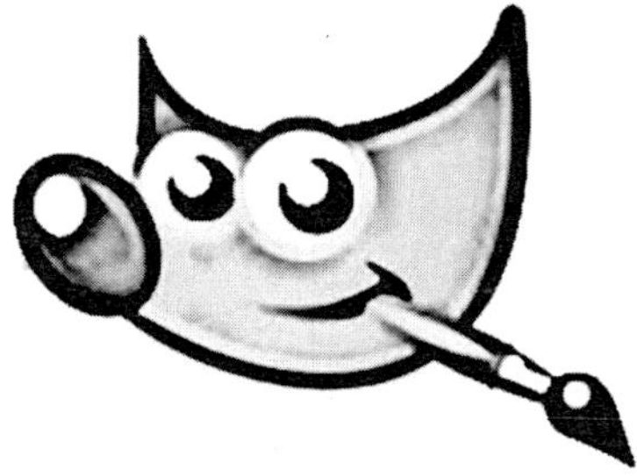

Key Features of GIMP

1. Image Editing: GIMP offers an array of tools for basic to advanced photo manipulation, such as cropping, resizing, color correction, and retouching.
2. Layers and Masks: Its support for layers and masks provides flexibility in composing and editing images non-destructively.
3. Color Adjustment: GIMP includes tools for adjusting color balance, brightness, contrast, hue, and saturation, giving users control over the aesthetic quality of images.
4. Filters and Effects: The software has a comprehensive set of filters and effects for adding stylization, texture, and artistic modifications.
5. Support for Different Formats: GIMP supports various file formats, including JPEG, PNG, GIF, TIFF, and even Photoshop PSD files.

Applications of GIMP

- Photography and Image Restoration: GIMP is extensively used for editing photographs, enhancing image quality, and restoring old or damaged photos.
- Graphic Design: It's a preferred tool for designing posters, banners, and web graphics due to its flexibility and extensive design tools.
- Digital Painting: GIMP's brush dynamics and custom brush tools make it suitable for digital artists to create illustrations and paintings.

Workflow with GIMP

1. Setting Up the Workspace: Customizing the workspace to include essential tools and shortcuts can streamline the editing process.
2. Image Import and Basic Adjustments: Upon importing an image, initial adjustments like resizing and cropping can set a foundation for further editing.
3. Layer Management: Layers enable users to stack different elements, allowing for complex compositions and easy adjustments.
4. Using Filters and Effects: Adding filters and adjusting levels can transform a simple image into a visually compelling piece.
5. Exporting Final Output: The final image can be exported in the desired format for sharing or further editing.

Overview of Blender

Blender is a comprehensive open-source 3D creation suite that supports all aspects of 3D development, including modeling, rigging, animation, simulation, rendering, compositing, and video editing. Unlike GIMP, which focuses on 2D graphics, Blender is ideal for 3D artists and animators.

Key Features of Blender

1. 3D Modeling and Sculpting: Blender provides tools for polygonal and sculpting-based 3D modeling, making it ideal for creating complex shapes and organic models.
2. Animation and Rigging: The software includes a sophisticated animation system, complete with rigging and weight painting tools, to bring characters to life.
3. Physics Simulation: Blender supports realistic simulations for physics-based effects like fluids, smoke, particles, and cloth.

4. Rendering: With its Cycles and Eevee render engines, Blender can create photorealistic or stylized images and animations.
5. Video Editing and Compositing: Blender has a built-in Video Sequence Editor (VSE) for video editing and compositing, making it a multi-purpose tool for video and 3D work.

Applications of Blender

- 3D Animation and Game Development: Blender is widely used for creating animated films, 3D video games, and VR experiences.
- Architectural Visualization: It's a go-to tool for architects and designers to create photorealistic 3D models and renderings of buildings.
- Product Design and Visualization: Companies use Blender for creating product prototypes and virtual renderings to test and showcase new products.

Workflow with Blender

1. Interface Navigation: Blender's user interface may initially appear complex, but understanding its layout and tools is essential for efficient 3D work.
2. Creating and Manipulating 3D Models: Using tools like extrude, loop cut, and bevel, users can create a wide variety of 3D shapes and objects.
3. Texturing and Shading: Texturing adds realism, while shading provides material properties such as glossiness, transparency, and bump.
4. Animating and Keyframing: With Blender's timeline and keyframe tools, animators can create movement and character actions.
5. Rendering the Final Product: The rendering process brings together lighting, shading, and animation to produce the final image or animation.

Integrating GIMP and Blender

Both GIMP and Blender can be used in tandem to enhance workflows in digital content creation. For instance, a designer might use GIMP to create textures and then import them into Blender for application on 3D models. Similarly, post-render adjustments, such as color correction and compositing, can be performed in GIMP.

Sample Workflow for 3D Texturing

1. Designing Textures in GIMP: Start by creating high-resolution textures in GIMP, using its layering and masking tools for intricate details.
2. Applying Textures in Blender: Import the texture into Blender and apply it to the 3D model, adjusting UV mapping to fit the texture correctly.

3. Rendering and Final Touches: After rendering, use GIMP to make final adjustments, such as color balancing, contrast enhancement, and minor touch-ups.

Advantages and Limitations of Open-Source Tools

- Advantages: Open-source software like GIMP and Blender are cost-effective, highly customizable, and supported by large communities, providing ample resources and plugins.
- Limitations: They may lack certain proprietary features and integrations, and the learning curve can be steep for beginners.

Conclusion

The application of GIMP and Blender exemplifies the potential of open-source software in creative fields. From digital art to 3D animations, these tools have made it possible for artists, designers, and engineers to create high-quality work without incurring software costs. By understanding the workflows and best practices for each software, users can maximize their creative output and achieve professional results. With continued advancements in open-source technology, GIMP and Blender are likely to remain vital resources for digital creators worldwide.

23

Open Broadcasting Software

Introduction to Open Broadcasting Software (OBS)

Open Broadcasting Software, commonly known as OBS, is an open-source platform designed for video recording and live streaming. It has become the software of choice for content creators, gamers, educators, and businesses due to its flexibility, powerful features, and ease of use. OBS is cross-platform, available on Windows, macOS, and Linux, and supports a range of applications from streaming live gameplay to creating professional-looking video presentations and tutorials.

Key Features of OBS

- Customizable Interface: OBS provides a highly customizable interface that allows users to arrange different elements like audio controls, scenes, sources, and transitions. This helps users create a personalized workspace that suits their specific broadcasting needs.
- Multiple Source Types: OBS allows users to incorporate a variety of sources, such as screen captures, webcams, images, text, browser windows, and more, making it ideal for producing complex and visually engaging streams.

- Real-Time Audio and Video Capture: OBS offers real-time capture and mixing, giving users control over multiple video and audio sources in one interface. Users can modify audio quality, add filters, and use visual effects in real-time.
- High-Quality Recording and Encoding Options: OBS supports high-definition (HD) streaming and recording, along with various encoding options. It utilizes x264, H.264, and hardware encoding, balancing quality with efficient CPU usage.
- Scene Management and Transitions: OBS allows users to create multiple scenes, each containing its own unique configuration of sources. Users can switch between scenes using smooth transitions, creating a dynamic viewing experience.
- Virtual Camera: OBS's virtual camera feature allows users to use OBS as a virtual camera source in video conferencing applications. This feature is particularly useful for educators, marketers, and others who want to create enhanced presentations in real-time.

Setting Up OBS: A Step-by-Step Guide

- Installation and Initial Configuration: Download OBS from its official website, and follow the installation instructions for your operating system. Upon the first launch, OBS provides an auto-configuration wizard to help users set up optimal settings based on their hardware and streaming preferences.
- Creating and Managing Scenes: A scene in OBS is a collection of sources, including images, text, screen captures, and webcams, arranged to produce a specific visual layout. To create a scene, users can click the '+' button in the Scenes panel, then add sources to bring in the desired content.
- Adding and Customizing Sources: Adding sources is as simple as selecting them from the Sources panel and positioning them within the scene. OBS offers fine control over positioning, layering, and sizing, allowing for precise visual arrangements.
- Configuring Audio Sources: OBS's Audio Mixer panel provides control over each audio source's volume, balance, and mute settings. Users can add audio filters, including noise suppression, gain control, and VST plugins for more advanced audio editing.

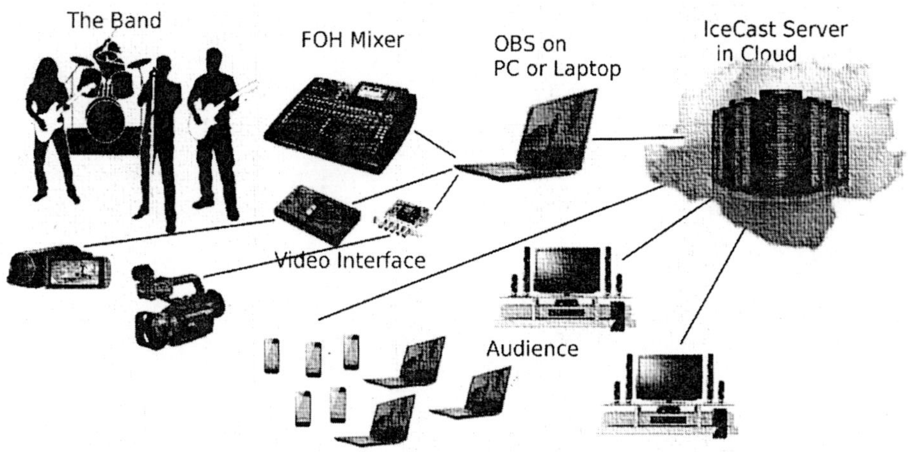

Steps in Open Broadcasting Software (OBS)

Advanced Techniques in OBS

- Encoding Settings for Optimized Performance: OBS provides various encoding options, allowing users to balance video quality with system performance. x264 and H.264 encoding are standard for most streaming platforms, with advanced users taking advantage of hardware encoders like NVENC and AMF to offload processing from the CPU to the GPU.
- Using Filters and Effects: OBS supports a range of filters for video and audio sources, such as chroma key (for green screen effects), color correction, and noise suppression. These filters help in enhancing the visual and audio quality of the broadcast.
- Setting Up Hotkeys: OBS enables hotkeys for switching scenes, starting or stopping streams, and adjusting audio levels, allowing for seamless control during live broadcasts without needing to navigate the interface.
- Recording and Streaming Simultaneously: OBS allows users to record locally while streaming live, which is beneficial for creating archived content or for further editing and distribution on other platforms.

Integration with Third-Party Plugins and Extensions

- OBS Studio Plugins: OBS has a vibrant plugin ecosystem that enhances its core functionality. Plugins like Streamlabs and Voicemeeter expand OBS's capabilities, allowing for in-depth sound routing, animated alerts overlays, and advanced scene transitions.
- Virtual Audio and Visual Effects Plugins: Plugins like VST audio plugins and the Virtual Camera plugin empower creators to customize

audio profiles and apply visual effects, creating more professional and immersive broadcasts.

- Stream Elements and Overlay Integrations: Many content creators use overlays and alerts to engage with their audiences. Services like Stream Elements and Streamlabs provide overlays, animations, and real-time audience interactions, which can be integrated seamlessly with OBS.

Best Practices for Effective Streaming with OBS

- Preparing Scenes and Sources in Advance: Pre-configuring scenes and organizing sources before a live session helps ensure smooth transitions and a consistent viewing experience.
- Optimizing System Resources: OBS can be resource-intensive, particularly on older machines. It's important to monitor CPU and GPU usage and to lower settings like output resolution or frame rate if necessary to prevent lag or dropped frames.
- Testing Streams Before Going Live: A pre-broadcast test helps to identify any potential issues with audio sync, source layout, or stream quality. OBS allows for unlisted testing on platforms like YouTube, providing an opportunity for final adjustments.
- Engaging with the Audience: OBS provides tools like chat integration and interactive overlays that allow for real-time engagement, which can significantly enhance the viewer experience.

Applications of OBS Across Various Fields

- Gaming and Entertainment: OBS is widely used by gamers for live-streaming gameplay on platforms like Twitch and YouTube. The software's performance optimization and high-quality recording options make it ideal for capturing fast-paced, high-definition content.
- Education and Training: Educators use OBS for online lectures, screen recording, and interactive presentations. Features like the virtual camera enable teachers to integrate OBS content directly into video conferencing software, creating more dynamic instructional materials.
- Corporate and Marketing Uses: In corporate settings, OBS is used for live events, webinars, and product demonstrations. The software's compatibility with platforms like Zoom and Teams, along with its customizable branding options, make it a cost-effective solution for businesses.
- Social Media Content Creation: OBS is popular among vloggers and influencers who create live or recorded content for social media. The

ability to integrate multiple sources and scenes allows for visually diverse and interactive content tailored for platforms like Instagram and Facebook.

Challenges and Limitations of OBS

While OBS is powerful, it does have some challenges:

- Resource Demands: OBS can be resource-heavy, especially during high-quality streams, which can affect system performance on lower-spec computers.
- Learning Curve: While intuitive for tech-savvy users, OBS has a learning curve for beginners, particularly in the areas of scene management and encoding settings.
- Limited Built-In Support for Advanced Editing: OBS is designed for live recording and streaming rather than post-production editing, which requires separate software for tasks like detailed video cutting and effects.

Future Directions and Community Contributions

As an open-source platform, OBS is continuously evolving. The OBS community actively develops plugins and customizations that expand its functionalities. Features such as AI-driven filters, advanced automation tools, and mobile support are anticipated to enhance OBS further. Community contributions through GitHub play a significant role in maintaining and advancing OBS, making it an evolving tool that keeps pace with the dynamic needs of modern content creators.

Conclusion

OBS has transformed the content creation landscape, making high-quality broadcasting and streaming accessible to a broad audience. With its open-source nature, comprehensive features, and extensive plugin support, OBS has become an indispensable tool for anyone involved in online content creation. By mastering OBS, creators can leverage its powerful functionalities to produce professional-grade broadcasts that captivate and engage audiences across platforms.

24

Google Sketchup and Any Other

Introduction to 3D Modeling Software

The world of 3D modeling has transformed design, enabling users to create digital representations of objects, buildings, and landscapes with precision and detail. Whether for architecture, product design, or art, 3D modeling software has become invaluable for professionals and enthusiasts alike. This chapter explores Google SketchUp, one of the most accessible tools for 3D modeling, alongside other popular software, comparing features, strengths, and applications.

Key Concepts in 3D Modeling

- Meshes and Polygons: Basic building blocks of 3D models.
- Rendering and Texturing: Applying materials and lighting for realistic effects.
- Exporting and Compatibility: Sharing and integrating models across platforms.

Overview of Google SketchUp

Google SketchUp is known for its user-friendly interface and versatile tools, making it a popular choice for beginners and professionals.

History and Development

- Initially developed by @Last Software and later acquired by Google, SketchUp was designed to be intuitive and accessible.
- Today, it's owned by Trimble Inc., with multiple versions for varied user needs, including SketchUp Free, SketchUp Pro, and SketchUp Studio.

Key Features

1. Push/Pull Tool: Easily transforms 2D shapes into 3D models.
2. 3D Warehouse: A vast online library where users can download or share models.
3. Extensions and Plugins: Extends functionality for specialized modeling tasks.

4. Integration with Other Software: Compatibility with tools like AutoCAD and Revit.
5. Cloud-Based and Desktop Options: Flexibility for various workflows.

Applications of Google SketchUp

- Architecture: Designing buildings and interior spaces.
- Urban Planning: Visualizing cityscapes and landscape designs.
- Product Design: Prototyping products for testing and development.
- Gaming and Animation: Creating objects and settings for virtual environments.

Strengths and Limitations of Google SketchUp

- Strengths: Intuitive interface, versatile application, rich library, and extensive online resources.
- Limitations: Limited advanced rendering capabilities compared to other tools, especially for high-end graphic demands.

Alternatives to Google SketchUp

While SketchUp is powerful, other tools may offer specific features for unique needs. Here are some prominent alternatives, comparing their features with those of SketchUp.

1. Blender

Blender is a comprehensive, open-source 3D creation suite widely used for modeling, animation, and rendering.

- Strengths: High customization, robust rendering engine (Cycles), excellent for animation, and free.
- Limitations: Steeper learning curve, extensive tools that may overwhelm beginners.

- Best For: Advanced users who require detailed modeling, sculpting, and rendering capabilities.

2. Autodesk Fusion 360

A cloud-based 3D CAD, CAM, and CAE tool primarily used in product design and engineering.

- Strengths: Parametric design, simulation tools, integrates well with CNC workflows.
- Limitations: Subscription-based, intended mainly for engineering rather than creative arts.
- Best For: Engineers and product designers focused on creating functional prototypes.

3. Tinkercad

Developed by Autodesk, Tinkercad is a free, web-based modeling tool popular among beginners, especially children.

- Strengths: Simple, block-based interface, easy to learn, perfect for beginners.
- Limitations: Limited to basic shapes and designs, lacks advanced tools.
- Best For: Students, beginners, and hobbyists looking for an entry into 3D modeling.

4. Rhino 3D

Rhinoceros (or Rhino) is known for its precision and flexibility, especially in architecture and industrial design.

- Strengths: Excellent for detailed architectural and CAD models, includes NURBS modeling.
- Limitations: Higher price point, steeper learning curve.
- Best For: Architects and designers who need detailed and precise models, especially for organic and complex shapes. Comparison Table of 3D Modeling Software

Software	Cost	Best For	Notable Features
Google Sketch Up	Free/Pro	Architecture, design	3D Warehouse, Push/Pull tool
Blender	Free	Animation, game design	Advanced rendering, animation tools
Fusion 360	Subscription	Product engineering	Cloud-based, parametric design

Tinkercad	Free	Beginners, hobbyists	Browser-based, block-based interface
Rhino 3D	Paid	Architecture, CAD	NURBS modeling, precision tools

Selecting the Right Tool

Choosing the right 3D modeling software depends on the user's skill level, project requirements, and budget.

- For Beginners: Tinkercad and SketchUp offer straightforward tools ideal for entry-level users.
- For Intermediate Users: SketchUp Pro and Blender allow greater functionality and more detailed modeling.
- For Advanced Users: Rhino 3D and Fusion 360 provide high precision, especially useful in fields requiring technical specifications and parametric designs.

Practical Tips for Using 3D Modeling Software

1. Understand Basic Modeling Principles

Familiarize yourself with 3D spaces, axis orientations, and basic shapes. Learning about polygons, vertices, and textures can help in creating realistic models.

1. Leverage Community Resources

Most software offers tutorials, forums, and downloadable assets, such as SketchUp's 3D Warehouse and Blender's online community. These resources provide both guidance and inspiration.

2. Experiment with Rendering and Lighting

Adding realistic lighting and textures can greatly enhance the presentation of 3D models. SketchUp's V-Ray extension or Blender's Cycles renderer are excellent tools for these purposes.

Conclusion

Google SketchUp and other 3D modeling software are versatile tools that cater to different needs within the design world. From architecture to animation, the choice of software hinges on factors like ease of use, required precision, and user experience level. Experimenting with a few different tools can help users find the software that best suits their projects, maximizing their creativity and productivity.

25

Presentation of Produced Programme

Introduction

This chapter discusses the crucial final stage of any video or multimedia production: the presentation. A well-produced program needs to be presented effectively to captivate, inform, and engage the intended audience. This chapter explores various facets of presenting a produced program, including preparation, techniques for impactful presentation, adapting to the platform and audience, and post-presentation considerations.

Understanding the Importance of Presentation

- Setting the Tone: A well-presented program captures the essence and quality of the production. Presentation directly influences how audiences perceive the program's message, style, and relevance.
- Role in Communication: Presentation serves as the final link between the production team and the audience. It ensures the message is not only delivered but also absorbed and remembered.
- Branding and Identity: The presentation also reflects the program's identity and the reputation of the production team or organization, affecting brand perception and credibility.

Preparation for Presentation

- Technical Setup: Ensuring high-quality audio and video is essential for a smooth experience. Check all equipment such as microphones, projectors, cameras, and screens. This preparation should include testing audio levels, video playback, and internet connection for online presentations.
- Audience Analysis: Understand the demographic and expectations of your audience. Are they professionals looking for detailed data, or a general audience seeking an overview? Knowing the audience will help in tailoring content, examples, and language.
- Platform Consideration: Each platform has unique strengths and limitations. For instance:
- Television: Requires strict adherence to time and often incorporates scheduled breaks.
- Live Streaming: Offers real-time engagement but also demands careful monitoring of technical stability.
- In-Person: Allows for direct interaction, eye contact, and real-time feedback from the audience.
- Rehearsal and Dry Run: Conduct a full rehearsal to simulate the actual presentation. This practice helps identify any potential issues, assess the flow, and build confidence. Rehearsals also help presenters gauge pacing, transitions, and points of emphasis.

Crafting a Compelling Introduction

- Hooking the Audience: Start with an intriguing question, a shocking fact, or a powerful visual to immediately capture attention. For example, starting with, "Did you know that one-third of the world's food is wasted?" can grab the audience's curiosity on a program about food sustainability.
- Setting Context: Within the first few moments, provide a clear and concise overview of the program's purpose. This orientation helps the audience understand what to expect and builds interest.
- Establishing Credibility: Introduce the team behind the program, emphasizing their expertise and dedication. A brief mention of the research, fieldwork, or unique production elements involved can establish the program's reliability and authority.

Guiding the Audience Through the Program

- Structuring the Presentation: Organize the program into clear sections, using transitional phrases to guide viewers from one point to the next. Avoid cluttered transitions; smooth, concise transitions help maintain flow.
- Storytelling Techniques: Employ a narrative arc if the content allows it, with a clear beginning, middle, and end. Techniques like foreshadowing key points or building suspense can keep the audience engaged.
- Visual and Audio Aids: Use graphics, animations, and background music to reinforce key messages or emphasize emotions. Ensure visuals are high quality, as low-resolution images can reduce perceived production value.
- Interactive Elements: When possible, include interactive elements like Q&A sessions, audience polls, or live comments, especially in online formats. This engagement can boost attention, participation, and retention of the content.

Adapting for Different Platforms

- In-Person Presentation:
- Body Language and Eye Contact: These are crucial for establishing a connection. Make eye contact with different sections of the audience, and use gestures to emphasize points.
- Microphone Use: If using a microphone, maintain consistent distance and volume. Handheld microphones require controlled movement to avoid noise.
- Online or Hybrid Presentation:
- Screen Layout: Optimize the layout for online platforms. For example, minimize the video window during screen sharing to focus on visuals.
- Audience Engagement Features: Use chat features, virtual hand-raises, or reactions to gauge engagement.
- Technical Monitoring: It's essential to monitor internet speed, lag issues, and audio clarity throughout the session.
- Broadcast and Streaming:
- Time Constraints: Network TV and online platforms often impose time limits. Be prepared to condense or expand content as needed, keeping core points intact.
- Ad Integration: If the platform has ads, plan transitions that ensure seamless resumption after breaks.

Overcoming Presentation Challenges

- Technical Difficulties: Always have a contingency plan. For instance, prepare a backup presentation on a USB, or have an alternate internet connection ready for online presentations.
- Audience Distractions: For in-person presentations, re-engage attention by changing the tone, asking a question, or introducing a new visual element.
- Time Management: Rehearse to ensure all points are covered within the allotted time. If necessary, identify secondary points that can be trimmed without losing the core message.

Concluding the Presentation Effectively

- Summing Up Key Points: Use concise language to reiterate main points. For example, "To recap, our program has highlighted the causes, effects, and solutions to climate change."
- Call to Action: Depending on the content, encourage the audience to take a next step—whether it's visiting a website, reading further, or participating in a survey. Calls to action are particularly effective for topics related to awareness and advocacy.
- End with Impact: Conclude with a memorable statement, perhaps a quote, a compelling image, or a thought-provoking question. This final moment should leave the audience with a lasting impression and ideally motivate further thought or action.

Significance of Presentation

- Captures audience attention
- Enhances content impact
- Builds brand identity
- Drives audience retention

Challenges in Presentation

- Balancing quality and cost
- Adapting to diverse audience preferences
- Navigating platform algorithms

Comparative Analysis: Television vs. Social Media

Aspect	Television	Social Media
Audience Reach	Broad but less interactive	Targeted and highly interactive
Format	Structured and time-bound	Flexible and often informal
Monetization	Ad-based and subscription models	Ad revenue, sponsored posts
Feedback	Delayed and indirect	Immediate and direct
Technology Use	High-end production tools	Accessible, app-based tools

Techniques for Enhancing Presentation across Platforms

Storytelling Techniques

- Use of compelling narratives to retain audience interest.
- Emphasis on emotional and relatable content.

Visual and Audio Design

- Consistent branding elements (e.g., logos, colors).
- High-quality soundtracks and effects to elevate the content.

Audience Engagement Strategies

- Incorporating polls, Q&A, and live sessions for real-time interaction.
- Regularly analyzing metrics to refine presentation styles.

Future Trends in Programme Presentation

- **Integration of Artificial Intelligence**
  - AI tools for personalization, automatic editing, and content recommendations.
- **Interactive and Immersive Content**
  - Adoption of virtual reality (VR) and augmented reality (AR) for enhanced viewer experiences.
- **Cross-Platform Integration**
  - Seamless sharing and viewing experiences between television and social media platforms.

26

Creating Videos Using Artificial Intelligence

The rise of Artificial Intelligence (AI) has significantly transformed video creation processes, enabling faster production, improved quality, and increased personalization. This chapter delves into the role of AI in video creation, exploring its applications, tools, techniques, and implications. It highlights how AI-powered tools enhance scripting, editing, animation, and special effects, and provides insights into real-world applications across industries such as marketing, education, entertainment, and more. The use of Artificial Intelligence (AI) in video production has transformed the landscape of media creation. From automating mundane tasks to enhancing creativity through intelligent algorithms, AI offers innovative tools that streamline the video production process. This chapter provides a comprehensive bibliography of key texts, research papers, articles, and online resources that delve into various aspects of creating videos using AI. These resources will be instrumental for scholars, practitioners, and enthusiasts seeking to deepen their understanding of AI in video production.

The integration of AI into video production has disrupted traditional workflows, making video creation more accessible, efficient, and innovative. From automating mundane tasks to introducing new creative possibilities, AI is shaping the future of visual storytelling.

What is AI in Video Creation?

AI in video creation refers to the use of machine learning algorithms and computer vision techniques to automate or augment tasks such as scriptwriting, editing, animation, and content personalization.

Why AI in Video Creation?

- Speeds up production processes
- Enhances creativity with advanced tools
- Enables hyper-personalization for targeted audiences
- Reduces production costs

AI Applications in Video Creation

Scriptwriting and Storyboarding

AI tools like Jasper and ChatGPT assist in generating scripts based on input keywords, themes, or objectives. They can also help in creating visual storyboards by mapping scenes to narratives.

Video Editing

AI-driven editing software like Adobe Premiere Pro with Sensei and Descript automates tasks like:

- Cutting and sequencing clips
- Enhancing audio quality
- Adding transitions and effects
- Color grading

Animation and CGI

AI tools like Runway ML and DeepMotion simplify the creation of animations and computer-generated imagery (CGI), enabling real-time motion capture and character generation.

Voiceovers and Subtitles

Natural language processing (NLP) and text-to-speech (TTS) systems, such as Murf.ai and ElevenLabs, generate realistic voiceovers. AI can also automate subtitle generation with high accuracy.

Personalization and Content Optimization

AI algorithms analyze viewer preferences to create personalized video recommendations, as seen in platforms like Netflix. Marketers use tools like Vidyard to tailor video content for individual users.

Special Effects and Augmented Reality (AR)

AI enhances special effects and integrates AR into videos, enabling immersive experiences. For example, platforms like Effect House allow creators to design AR effects.

AI-Powered Tools for Video Creation

Tool	Functionality	Key Features
Runway ML	Video editing and AI effects	Green screen, text-to-video, rotoscoping
Descript	Video and audio editing	Text-based editing, transcription
Synthesia	AI video generation with avatars	Multi-language support, customizable avatars
Pictory.ai	Automated video creation from text	Highlights detection, captions
Adobe Premiere Pro (Sensei)	Advanced editing with AI capabilities	Scene detection, auto-reframe

Real-World Applications of AI in Video Creation

Marketing and Advertising

AI allows marketers to produce dynamic, personalized ads that adapt to user behavior, increasing engagement and conversions.

Education and Training

AI-generated videos are used for creating interactive educational content, virtual tutorials, and corporate training modules.

Entertainment and Media

In the entertainment industry, AI accelerates post-production processes, enhances visual effects, and generates realistic digital characters.

Social Media Content Creation

AI tools enable influencers and businesses to create quick, engaging videos optimized for platforms like Instagram, TikTok, and YouTube.

Advantages of Using AI in Video Creation

Efficiency and Speed

AI automates time-consuming tasks, allowing creators to focus on storytelling and strategy.

Cost-Effectiveness

AI reduces the need for large production teams, minimizing costs associated with traditional video production.

Enhanced Creativity

AI tools expand creative possibilities, offering features like real-time animation and generative art.

Accessibility

Even non-technical users can create high-quality videos using AI-powered platforms with intuitive interfaces.

Challenges and Ethical Considerations

Over-Reliance on Automation

Excessive reliance on AI may stifle originality and human creativity.

Deep fakes and Misinformation

AI-generated videos can be misused to create deceptive content, posing ethical concerns.

Data Privacy

Personalized video creation often relies on user data, raising privacy and security issues.

Skill Gap

Adoption of AI tools may create a gap for professionals unfamiliar with advanced technologies.

Future Trends in AI-Driven Video Creation

Text-to-Video Technologies

Advanced tools like Meta's Make-A-Video are pushing the boundaries of generating realistic videos directly from textual descriptions.

Interactive and Immersive Videos

AI-powered interactive videos with features like branching storylines and real-time AR integration are gaining traction.

Real-Time Video Production

Advancements in real-time rendering and motion capture will revolutionize live video production.

Conclusion

AI is revolutionizing video creation, enabling creators to produce high-quality, engaging, and personalized content efficiently. While challenges such as ethical considerations and data privacy remain, the benefits far outweigh the drawbacks. As AI technologies evolve, they will continue to reshape the landscape of video production, making it more innovative, accessible, and impactful.

Bibliography

Adams, T., & McCormick, R. (2021). Effective Audio Recording Techniques for Mobile Devices. Journal of Digital Media, 12(3), 45-62.

Adobe Audition for Precise Audio Sync. Adobe, Inc. (2024). Retrieved from https://www.adobe.com/

Adobe Inc. (2022). Adobe Premiere Rush User Guide. Available at: https://helpx.adobe.com.

Adobe Inc. (2023). Adobe After Effects User Guide. Retrieved. from https://helpx.adobe.com/after-effects/user-guide html

Adobe Inc. (2023). Adobe Premiere Pro User Guide. Retrieved from https://helpx.adobe.com/in/premiere-pro/user-guide.html

Adobe Inc. (2023). Adobe Premiere Rush User Guide. Retrieved from https://helpx.adobe.com/premiere-rush/user-guide.html.

Adobe Inc. (2023). Adobe Stock User Guide. Retrieved from https://helpx.adobe.com/in/stock/user-guide.html

Adobe Inc. (2024). Creating Text Effects in Photoshop for Video. Available at: https://www.adobe.com/photoshop

Adobe Stock. (2024). Creating a cohesive video project with stock elements. Retrieved from Adobe Stock

Adobe Systems Inc. (2022). AI in Creative Cloud: Enhancing Digital Content Production.

Adobe Systems Incorporated. (2023). Understanding Codecs in Video Editing. Adobe Knowledge Base.

Adobe Systems Incorporated. Adobe Premiere Pro User Guide. Retrieved from https://helpx.adobe.com

Adobe. (2020). After Effects User Guide. Retrieved from https://helpx.adobe.com/after-effects/user-guide.html

Adobe. (2021). "Working with Proxies in Premiere Pro." Adobe Help Center. https://helpx.adobe.com/premiere-pro/using/proxies.html

Adobe. (2023). Photoshop for Video: A Guide to Using Adobe Photoshop in Video Production. Retrieved from Adobe Help Center.

Adobe.(2024)."Adobe Sensei: AI for Video Production." [Adobe Blog] https://blog.adobe.com/en/publish/2024/adobe-sensei-ai-video

Allen, R. (2022). Export Process in Video Editing: A Step-by-Step Guide. Multimedia Publishing.

Alten, S. R. Audio in Media. Wadsworth Publishing, 2010.

Anchor by Spotify. (2021). Podcasting Made Easy: A Guide to Creating and Publishing Audio Content with Your Smartphone. Anchor Publications.

Apple Inc. (2023). iMovie for iOS. Retrieved from https://www.apple.com/imovie/

Apple. (2020). "Editing Proxies in Final Cut Pro X." Apple Support. https://support.apple.com/guide/final-cut-pro/create-proxies-ver590cb4e2/mac

Apple. (2020). Final Cut Pro X User Guide. Retrieved from https://support.apple.com/guide/final-cut-pro/welcome/mac

Ardour Team. (2023). Ardour: The Digital Audio Workstation. Retrieved from https://ardour.org/

Armitage, A. (2020). Mastering DSLR Photography: A Complete Guide to Digital SLR Photography and Editing (2nd ed.). Amherst Media.

Ascher, S., & Pincus, E. (2013). The Filmmaker's Handbook: A Comprehensive Guide for the Digital Age. New York: Penguin.

Audacity Development Team. Audacity User Manual. Retrieved from https://www.audacityteam.org/manual

Audacity Team. (2023). Audacity: Free, Open Source, Cross-Platform Audio Software. Retrieved from https://www.audacityteam.org/

Autodesk Inc. (2023). Autodesk Maya Documentation. Retrieved from [Autodesk Maya] https://www.autodesk.com/products/maya/overview

Avid Technology. (2023). Media Composer Documentation.

Avid. (2021). Media Composer 2021 User Guide. Retrieved from https://www.avid.com/media-composer

Avid. (2022). "Avid Media Composer Proxy Workflow." Avid Knowledge Base. https://avid.force. com/ pkb/articles/ How_to/ Avid- Media-Composer-Proxy-Workflow

Bamford, S. (2014). Digital Photography: A Beginner's Guide to Taking Professional Photos and Using DSLR Cameras. Independently published.

Bengio, Y., Goodfellow, I., & Courville, A. (2016). Deep Learning. MIT Press.

Bill, A. (2019). Motion Graphics: Principles and Practice. Routledge.

Blackmagic Design. (2023). DaVinci Resolve 18 User Manual. Blackmagic Design Documentation.

Blackmagic Design. DaVinci Resolve Manual. Retrieved from https://www.blackmagicdesign.com

Blackwell, D. (2021). "Open Source in Video Editing: Examining Blender and the Open Movie Project." Digital Arts Journal, 10(2), 85-102.

Blender Foundation, "Blender Video Editing," accessed https://www.blender.org/

Blender Foundation. (2021). Blender 3D User Manual. Retrieved from https://docs.blender.org/manual/en/latest/

Blender Foundation. (2023). "Blender Video Editing." Retrieved from https://www.blender.org/

Blender Foundation. (2023). Blender 3D Modeling Documentation. Retrieved from [Blender] https://www.blender.org/

Blender Foundation. (2023). Blender Documentation: Video Editing and 3D Animation. Retrieved from https://www.blender.org/

Blender Foundation. (2023). Blender Graphics and Video Library. Retrieved from https://www.blender.org/

Blender Foundation. (2023). Blender Video Editing and Animation. Retrieved from https://www.blender.org/

Blum, R. (2021). Digital Audio Essentials: A Comprehensive Guide to Digital Audio for Music, Video, Film, and Mobile Devices. O'Reilly Media.

Bojanowski, P., & Douze, M. (2021). AI Tools for Video Synthesis and Animation. ACM Computing Surveys, 54(5), 112-136.

Bordwell, D., & Thompson, K. (2019). Film Art: An Introduction. McGraw-Hill Education.

Broll, J. (2022). "Editing on the Move: The Best Mobile Video Editing Apps." Videomaker Magazine, 41(4), 67-72.

Brown, B. (2016). Cinematography: Theory and Practice: Image Making for Cinematographers, Directors, and Videographers (3rd ed.). Routledge.

Brown, B. (2018). The Filmmaker's Guide to Digital Editing. Routledge.
Brown, L. (2021). Photoshop Basics for Video Editors. VideoMaker.
Busche, A. (2014). Setting up your shots: Great camera moves every filmmaker should know (2nd ed.). Michael Wiese Productions.
Canva. (2024). Designing social media graphics with templates. Retrieved from Canva
Chen, D., & Kim, S. (2022). "Waveform Free and the Democratization of Audio Production Tools." Journal of Open Source Audio Software, 6(3), 45-63.
Cheng, Y., & Liu, H. (2021). "AI in Video Production: A Review." Journal of Visual Communication and Image Representation, 77, 103-115.
Chien, B. (2017). Film Editing: History, Theory and Practice. Routledge.
Christiansen, M. (2020). Designing Motion Graphics for Video Titles. Adobe Press.
Cockos Incorporated. (2023). REAPER User Guide. Retrieved from https://www.reaper.fm/userguide.php
Creative Cloud Team. (2022). Resolution and Aspect Ratios in Modern Video Production. Adobe Blog.
Curtin, D. P. (2020). Digital Photography: A Hands-on Introduction to Photography and Photo-Editing (4th ed.). Delmar Cengage Learning.
DaVinci Resolve. (2022). "Working with Proxy Media." Blackmagic Design. https://www.blackmagicdesign.com
Digital Media Editing, (2021). Comprehensive Guide to Adobe Premiere Pro Export Settings. Pearson Education.
Dolby On App. (n.d.). Retrieved from Dolby On
Dorsey, J. (2016). Motion Graphics: Design and Practice. Fairchild Books.
Dufresne, B., & Silva, C. (2019). Advanced Editing Techniques in Premiere Pro and Final Cut Pro X. Packt Publishing.
Envato Elements. (2024). Using templates and stock footage for quick project turnaround. Retrieved from Envato
Evening, M. (2020). Adobe Photoshop for Photographers (5th ed.). Focal Press.
Fernandez, S., & Jansen, H. (2020). "AI-Powered Video Creation: Trends and Best Practices." In Proceedings of the International Conference on Digital Media and Communication (pp. 112-118). ACM.
Ferrite Recording Studio. (n.d.). Retrieved from Ferrite Recording Studio
Fisher, J. (2020). Audio Production Tips and Tricks for Content Creators. Taylor & Francis.
Focusrite, "Getting Started with Your Scarlett Interface," Focusrite Website
Freepik. (2024). Graphics for any occasion: How to use icons and vectors effectively. Retrieved from Freepik
Fuchs, S. (2021). Synchronized Sound: The Art of Audio Editing for Video. New York: Routledge.
Gent, E. (2020). "The Best Open Source 3D Modeling Software." PCMag. Retrieved from [PCMag] https://www.pcmag.com
Gervais, R. Home Recording Studio: Build It Like the Pros. Cengage Learning, 2011.
Gibson, R. (2021). A Guide to DAWs: Reaper for Independent Music Production. Sound Design Quarterly, 9(2), 99-117.
Gimenez, P., & Busselle, M. (2016). The DSLR Filmmaker's Handbook: Real-World Production Techniques (2nd ed.). Wiley.
Gonzalez, M., & Lee, J. (2020). "The Role of Artificial Intelligence in Content Creation: Opportunities and Challenges." International Journal of Creative Media Studies, 4(2), 45-62.

Goodfellow, I., & Pouget-Abadie, J. (2020). Generative Adversarial Networks (GANs) in Video Synthesis. MIT Press.

Google LLC. (2023). Google SketchUp Help Center. Retrieved from [Google SketchUp] https://www.sketchup.com/

Green, E. (2021). Efficient Storage Solutions for Video Files. Video Production Monthly.

Gustavson, T. (2009). Camera: A History of Photography from Daguerreotype to Digital. Sterling Innovation.

Harrington, R. (2021). Photoshop for Video Editors: Using Photoshop and After Effects. Peachpit Press.

Harris, S., & Higgins, L. (2021). Non-linear Editing: Theory and Techniques for Video Editors. Focal Press.

Healy, C., & Li, R. (2022). "Professional-Grade Editing with DaVinci Resolve Free Version." Creative Media Journal, 19(4), 345-360.

Hedgecoe, J. (2005). The New Manual of Photography. DK Publishing.

Holman, T. (2008). Sound for Film and Television. Focal Press.

Holman, T., & Plessas, A. (2020). Smartphone Filmmaking: Theory and Practice. Routledge.

Hunter, F., Biver, S., & Fuqua, P. (2021). Light: Science and Magic: An Introduction to Photographic Lighting (6th ed.). Focal Press.

InShot Inc. (2023). InShot App Tutorial and FAQ. Available at: https://www.inshot.com.

Izhaki, R. (2017). Mixing Audio: Concepts, Practices, and Tools. Routledge.

Jacobs, M. (2013). Photography Camera Basics 101: The Ultimate Guide for Understanding Digital Photography & Camera Settings for Beginners. Kindle Edition.

Johnson, D. (2021). Social Media Video Formats and Standards. Social Media Journal.

Johnson, M. (2023). Immersive Video: Exporting HDR, 360°, and VR Formats. AVA Publishing.

Kaufmann, A., & Schmidt, M. (2019). AI and the Future of Video Production. Technology and Creativity, 12(2), 35-50.

KDE, "Kdenlive: Free and Open Source Video Editor," accessed https://kdenlive.org/

KDE. (2023). "Kdenlive: Free and Open Source Video Editor." Retrieved from https://kdenlive.org/

Keating, P. (2014). Film and Video Production in the Digital Age: Theory and Practice. Routledge.

KineMaster Corporation (2023). KineMaster: Mobile Video Editing Made Easy. Available at: https://www.kinemaster.com.

Kinemaster Corporation. (2023). Kinemaster - Video Editor. Retrieved from https://www.kinemaster.com/

KineMaster Corporation. (2023). KineMaster User Guide. Retrieved from https://www.kinemaster.com.

Koehler, M. J. (2020). Artificial Intelligence in Education: A New Era of Learning. Springer.

Kose, D., & Zankov, A. (2022). "Audio-Visual Synchronization: Techniques and Best Practices in Mobile Editing." Journal of Multimedia and Digital Communication, 15(3), 210-224.

Kumar, R. (2020). Television Broadcasting: Techniques and Trends. New Delhi: Media Scholars.

Kumar, R., & Shukla, A. (2020). Video Editing with AI: Exploring Deep Learning and Computer Vision. Elsevier.

Langford, M., Fox, A., & Sawdon Smith, R. (2018). Langford's Basic Photography: The Guide for Serious Photographers (10th ed.). Routledge.

Lee, P. (2021). High-Resolution Video Editing in Practice. AVA Publishing.

Lester, P. M. (2013). Visual Communication: Images with Messages. Wadsworth Cengage Learning.

Li, H., & Lin, Y. (2021). "Introduction to 3D Modeling Software: A Comparative Study on Open Source and Commercial Options." Journal of Digital Media, 45(2), 58-76.

Li, S., & Zhang, M. (2020). AI-based Video Creation and Editing: Techniques, Tools, and Applications. Springer.

Liljeblad, J. (2023). Flowblade Video Editor Documentation. Retrieved from https://jliljebl.github.io/flowblade/

LMMS Community. (2023). LMMS: Free Music Production Software. Retrieved from https://lmms.io/

Martinez, A., & Kim, D. (2022). "Automating Video Editing with Machine Learning Techniques." Computers in Human Behavior, 121, 106-115.

Miller, S. (2020). The Art of Video Exporting. Creative Publishers.

Miller, T. (2022). File Management and Storage for Video Editors. Film School Press.

Mullen, T. (2022). The Complete Guide to Blender Video Editing. Blender Foundation. Retrieved from https://www.blender.org/support/tutorials

Murch, W. (2001). In the Blink of an Eye: A Perspective on Film Editing (2nd ed.). Silman-James Press.

Newman, M. & Block, M. (2015). The Filmmaker's Handbook: A Comprehensive Guide for the Digital Age (5th ed.). Plume.

OBS Project. (2023). OBS Studio: Open Broadcaster Software Documentation. Retrieved from [OBS Project]https://obsproject.com/

Ocenaudio. (2023). Ocenaudio: Easy, Fast, and Powerful Audio Editor. Retrieved from https://www.ocenaudio.com/

Official YouTube Creator Academy: https://www.youtube.com/creators

OpenAI. (2023). "DALL-E and Video Generation." https://openai.com/index/dall-e-3/

OpenShot Studios, LLC. (2023). "OpenShot Video Editor." Retrieved from https://www.openshot.org/

OpenShot Studios, LLC. OpenShot Video Editor Documentation. Retrieved from https://www.openshot.org

Ottman, R. (2022). Web Optimization Techniques for Video Export. Web Developer Magazine.

Patel, N. (2021). Social Media Marketing Essentials. New York: Digital Trends Press.

Patel, R., & Jansen, M. (2020). Audio-Visual Sync in Mobile Media Production: Challenges and Solutions. Media and Technology Journal.

Peres, M. R. (2013). The Focal Encyclopedia of Photography: Digital Imaging, Theory and Applications, History, and Science (4th ed.). Focal Press.

Pictory.ai: https://pictory.ai

Pohlmann, K. C. (2017). Principles of Digital Audio (6th ed.). McGraw-Hill Education.

Roberts, M. (2020). Mobile Video Editing: A Comprehensive Guide to Using Your Smartphone for Creative Projects. Boston: Cengage Learning.

Rocca, E., & Tan, B. (2021). "The Influence of AI on Video Production: A Case Study Approach." Media, Culture & Society, 43(4), 634-650.

Rumsey, F., & McCormick, T. (2014). Sound and Recording: An Introduction (7th ed.). Routledge.

Runway ML: https://runwayml.com

Saltz, I. (2021). Typography Essentials: 100 Design Principles for Working with Type. Rockport Publishers.

Savage, S. (2011). The Art of Digital Audio Recording: A Practical Guide for Home and Studio. Oxford University Press.

Seymour, P. (2021). The AI Revolution in Film and Video: Understanding the Technology. MIT Press.

Shao, M., & Wang, Z. (2021). AI-Powered Video Creation Tools: A Practical Guide. Oxford University Press.

Shapiro, A. (2019). AI and the Future of Video Production. Routledge.

Sharma, R. (2021). AI and Media: Redefining Video Content Creation. New York: TechPress.

Shotcut Documentation Team. (2023). Shotcut User Manual and FAQ. Retrieved from https://shotcut.org/FAQ/

Shotcut. (2023). "Shotcut: Free, Open Source, Cross-Platform Video Editor." Retrieved from https://shotcut.org/

Smartphone Audio Tips. (2022). Audio Engineering Society Journal, 30(6), 123-129.

Smith, A. (2022). Engaging Audiences in the Digital Age. London: Media Futures Publishing.

Smith, J. (2020). The Complete Guide to Video Formats and Codecs. TechMedia Press.

Smith, J. (2022). The Mobile Video Editor's Guide: Synchronizing Audio with Video. Creative Media Publishing.

Thomas, R., & Smith, R. (2015). The Art of 3D Computer Animation and Effects. Wiley.

Tracktion Corporation. (2023). Waveform Free Manual. Retrieved from https://www.tracktion.com

Vaswani, A., Shazeer, N., Parmar, N., Uszkoreit, J., Jones, L., Gomez, A., Kaiser, Ł., & Polosukhin, I. (2017). Attention is All You Need: A New Architecture for AI Video Generation. NeurIPS, 30, 5998-6008.

Veed.io.(2023)."How to Use AI to Edit Videos." [Veed.io] https://www.veed.io/

Video Production Guide. (2022). Optimizing Bitrate and Compression in Video Editing. Media Insights.

Videvo. (2024). Maximizing stock video usage in digital projects. Retrieved from Videvo

Villalobos, J., & Harper, B. (2023). "The Evolution of Open Source Video Editing Tools and Their Impact on Independent Media Production." Journal of Open Source Practices, 7(1), 20-35.

Visual Effects Society. (2022). Text Integration in Video: Best Practices. VFX Handbook.

VoiceQ: Advanced Dubbing and ADR Software. (2024). Retrieved from https://voiceq.com/

Wang, T. (2022). Video Production with Artificial Intelligence: Techniques and Applications. CRC Press.

Webster, P. (2019). Smartphone Audio: Achieving High-Quality Voice Recording on Mobile Devices. Routledge.

White, P. (2003). Basic Mixing Techniques. Sanctuary Publishing Ltd.

Williams, L. (2023). Compression Techniques in High-Resolution Video. Motion Picture Press.

Williams, L., & Smith, A. (2021). "Comparative Analysis of Open Source Video Editing Software: Usability and Performance." International Journal of Software Studies, 14(3), 155-167.

Young, P. (2021). "Exploring Mobile Apps for Video Editing: From Concept to Creation." Journal of Digital Media Technology.

YouTube (2024). Recommended upload encoding settings. Available at: https://support.google.com/youtube.

YouTube Help Center. (2022). Guidelines for HDR and 360° Video Uploads. YouTube Documentation.

Yun, X., & Lee, J. (2022). Transforming Video Creation with AI: Innovations in Automation and Creativity. Journal of Multimedia and AI, 41(3), 76-92.

Zhang, L., & Lee, T. (2020). Machine Learning for Visual Media. Boston: MIT Press.

Zhang, X., & Patel, R. (2021). "Innovative Video Editing Techniques Using AI." In Proceedings of the International Conference on Multimedia and Expo (ICME) (pp. 201-206). IEEE.

Zhao, B., & Wei, Y. (2020). Generative AI in Video: Algorithms and Applications. Frontiers in Artificial Intelligence, 3, 121-137.

Index